Stephen Dalton
THE MIRACLE OF
Flight

P1 (*overleaf*) A barn owl (*Tyto alba*)
carrying a young rat to its hungry young

Stephen Dalton

THE MIRACLE OF
Flight

McGraw-Hill Book Company

NEW YORK   ST. LOUIS   SAN FRANCISCO   TORONTO

Library of Congress Cataloging in Publication Data

DALTON, STEPHEN. The miracle of flight.

1. Flight.   2. Aerodynamics.   I. Title.
TL570.D314   591.1'852   76-52448
ISBN 0-07-015207-1

Published in the United States by
McGraw-Hill Book Company, 1977.

This book was designed and produced by
Blacker Calmann Cooper Ltd. London.

Printed in Italy

Contents

P2 The brimstone (*Gonepteryx rhamni*) is a close relative of the cabbage white – its wings are half-way through their downstroke

Introduction

WINGS ARE AMONG THE MOST EXTRAORDINARY of the many complex structures that have evolved in the animal kingdom. Insects and birds between them make up over three-quarters of all land creatures, and it is their ability to fly which has made them so unusually successful and allowed them to develop into such a wide variety of species and live in so many climates. For man the mastery of mechanical flight probably represents his supreme technical achievement, and it has already extended enormously his domination of his environment.

In addition to its technical interest, flight is also one of the most beautiful and exhilarating activities to observe in insects and birds and to experience as a pilot or even as a passenger in an aircraft. This book attempts to show the different ways in which insects, birds and man have mastered the vast technological problems involved and to reveal in the photographs some of the beauty of flight. It is a large and absorbing field of study and inevitably a book of this size can do no more than give a very general idea of the facts, and the interest, of the subject.

The method followed is a step-by-step account of how winged objects, both animal and mechanical, fly. In describing this, the book traces the evolution of flight in

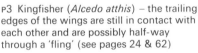

P3 Kingfisher (*Alcedo atthis*) – the trailing edges of the wings are still in contact with each other and are possibly half-way through a 'fling' (see pages 24 & 62)

chronological order from the first flying insects which rose from the swamps some 350 million years ago to the supersonic aircraft of the present day. As the book is only concerned with active sustained flight, it does not include flying fish, flying squirrels, flying snakes or other so-called 'flying' creatures which are only capable of remaining airborne for relatively short glides. Although bats are capable of true flight, they do not merit a separate chapter for their flight is broadly similar to that of birds, in spite of anatomical differences in their skeleton and wings.

Since, with the exception of balloons and rockets, almost all man's flying machines and certainly all living creatures depend on wings to support them in the air, the book concentrates on winged flight. The main purpose of the first chapter is to explain in simple terms the basic principles of flight. Although the subject of aerodynamics is highly specialized and, therefore, difficult for the ordinary person to comprehend, it is particularly important to have a grasp of the fundamentals in order to appreciate the many procedures that must be coordinated in order to achieve flight. This chapter, as well as all subsequent chapters, is illustrated with diagrams and photographs which, it is hoped, will clarify some of the more difficult points in the text and help the reader to understand the function, evolution and fascination of flight. It will also explain the reasons for the wide diversity of structures and mechanisms which exist both in the world of flying creatures and of flying machines. Finally, this chapter touches on the extraordinary discoveries recently made about insect and bird flight. Standard aerodynamic theories about steady airflow do not adequately account for many forms of animal flight; it was not until research at Cambridge, supported by the evidence of high-speed photography, showed that insects and birds make use of non-steady airflow that the differences between fixed-wing and flapping flight began to be understood.

After a brief retrospective look at the evolution of insect wings, the second chapter is concerned with insect flight and describes the fascinating, and hitherto unknown, mechanisms insects have developed for obtaining extra lift from the forces of the air. The third chapter, 'The Feathered Wing', is concerned with bird flight and, starting with the evolution from primitive feathered dinosaurs, describes the anatomy of the bird in relation to its flight, whether soaring, gliding, hovering or flapping.

In contrast to these chapters, the last two are devoted entirely to flight by man. In just a few generations manned flight has come all the way from the floundering attempts of the medieval tower-jumpers to the monumental achievement of the first powered, sustained and controlled flight by the Wright Brothers in 1903, which translated the mythical figures of Daedalus and Icarus into real life. Chapter 4 gives an account of the most important aeronautical developments and events from the earliest attempts up to the achievement of manned flight. The final chapter then goes on to explain how a conventional modern aeroplane flies and, by way of contrast with the 'micro-aerodynamics' of insect flight, includes man's latest technical achievements of supersonic flight and hang-gliding. As before, these chapters are illustrated by photographs that give a detailed picture of the insects, birds and machines themselves and diagrams that aim to clarify the technicalities involved. It is hoped that pictures and text will reveal something of the excitement of man's greatest achievement.

1
Fundamentals of flight

TO MOST OF US FLIGHT IS A MIRACLE; to primitive man it certainly must have seemed one, and it still remains mysterious even now that the physical laws governing it can be explained by the science of aerodynamics.

In his early attempts to fly man tried to imitate the flapping action of birds' wings, but he knew nothing about the properties of air or the forces that such wings harnessed, and so his efforts failed. Centuries later after experimenting with fixed wings, man started to raise himself from the ground, and at last he began to understand the nature of the medium into which he was venturing. He learnt as the insect and bird had learnt millions of years before him – by actual contact with the air's forces. Yet the ingenious methods he evolved appeared to have little in common with those of the bird whose performance he had tried to·emulate. Strangely enough, in the light of contemporary knowledge, it appears that the bird and the aeroplane function in much the same way.

Birds and insects have been on the wing for so long that their performance is instinctive and their 'knowledge' of the medium in which they fly is innate. For man, however, flying is a matter of intense effort and advanced technology. In order to appreciate fully the intrinsic qualities of flight in all its biological and mechanical diversity, it is essential to have some understanding of the principles of aerodynamics as they apply to the mass of air surrounding the surface of the earth, which is called the atmosphere. In broad terms, the science of aerodynamics is concerned with the motion of air; this, therefore, includes the displacement, speed and acceleration of its particles, as well as the physical forces exerted by air on all solid bodies, like birds or aircraft, travelling in it. This chapter attempts to explain these fundamentals in simple language. In doing so, many aspects of this complicated subject are covered step by step, as it is impossible to analyze them clearly in any other way; in the physical world which is being described, however, the occurrence of flight and the application of the laws of aerodynamics depend not only on a sequence of events, but also on the simultaneous coincidence of very many different movements or activities. This coincidence cannot be conveyed in words, but is apparent in the photographs, and, of course, even more in our observation of all types of flight by insects, birds or man-made machines.

P4 Biplanes flying in aerial display
P5 Blue tit (*Parus caeruleus*) carrying food

Whether upside down or the right way up,
the requirements of flight is a controllable
force directed upwards to overcome the
force of gravity

   The first serious attempt in Western history to analyze flight was made by Aristotle
about 350 BC. He maintained that an object, such as an arrow, could only continue
moving through the air so long as a force was applied to it, and that once this force was
withdrawn the object would stop. Furthermore, since he could not conceive how a force
could be conveyed from a distance, he claimed that the transmission of the object
required a force in physical contact with it. Aristotle thereby asserted that the air, or
atmosphere, actually perpetuated the flight, and he concluded that an object could not
be transmitted in a vacuum. An arrow, he argued, travelled through the air by being
pushed along by air rushing in to fill the vacuum behind it. He presumed that the air
sustained the flight of the arrow rather than retarded it. The Greek philosopher's views
on physics were not challenged until the fifteenth century when Leonardo da Vinci
(1452–1519) discarded his central thesis.
   Leonardo's assumption that the air was a resisting, rather than a sustaining, medium
was the first approximate explanation of the phenomenon of flight. It is on this basic
concept that the whole science of aerodynamics rests. However, his analysis of bird
flight was incorrect; he supposed that the motion of the flapping wings of birds caused
the air beneath to 'condense' and to behave like a rigid body on which the bird was
supported. On the same hypothesis, he explained that gliding flight, that is when the
wings are held motionless, depended on the relative motion between the wing and the

air, so that in a suitably strong wind the bird could be supported without beating its wings. A kind of condensation process resembling Leonardo's notion does in fact occur when wings are moved through the air, but only at extremely high speeds around the speed of sound, when the density of the air changes significantly.

One hundred years later Galileo's (1564–1642) suggestion that motion could exist on its own provided the death blow to Aristotle's theory. Like Leonardo, he recognized the action of air in resisting motion, and tried to define the precise way in which resistance changes with velocity. But a true scientific analysis of flight did not fall within man's grasp until Sir Isaac Newton (1642–1727) laid down the laws of motion and universal gravitation, which allowed his successors to study the properties of liquids and gases in a scientific manner. From that moment the secrets of the miracle of flight were gradually revealed to the world.

The chief requirement for flight is a controllable force directed upwards to overcome gravity. A balloon, a bullet and an aeroplane are all capable of flying in the broad sense of the word, but the reasons why they are supported in the air are entirely different. In each case the force of gravity, which urges their return to earth, is overcome. When a balloon floats in the air, it does so for the same reason that a cork floats in water: any object suspended in a liquid or gas experiences an upward thrust (or loss of weight) equal to the weight of the medium it displaces. This was first discovered by Archimedes in the third century BC. For example, a balloon which occupies 1,000 cubic feet of air is subjected to an upward thrust of about 80lb. (which is equal to the weight of 1,000 cubic feet of air); if the total weight of the balloon with its contents is less than 80lb., it floats upwards. Whether it floats or sinks will depend on the difference between the volume of air it displaces and its own total weight (D1).

The trouble with balloons is that they cannot be fully controlled and are entirely at the mercy of the wind; true flight requires that ascent, descent or movement in any desired direction should be possible irrespective of the wind. Airships, which are really rigid balloons with a means of propulsion, are capable of making headway in light winds, but, nevertheless, are inefficient and ponderous.

On the other hand, bullets and rockets are not supported by the air at all; they rely on their momentum or, in the case of burning rockets, thrust as well. They, together with all flying objects, obey Newton's first law of motion which states that a body stays at rest, or continues to move in a straight line and at a steady speed unless another force acts upon it to make it change its state. A bullet gains its initial thrust in the chamber of a rifle by an explosive charge, whereby it is rapidly accelerated to up to three times the speed of sound before it leaves the muzzle. In outer space the bullet would continue in a straight line and at a constant speed almost indefinitely, but in our atmosphere where other forces act upon it, its trajectory is shaped by gravity which steadily pulls it downwards, while its speed is slowed down by air resistance. The essential difference between projectiles and other flying objects is that projectiles do not rely on an atmosphere for their support; in fact, they become far more efficient in a vacuum where their momentum is not dissipated by the air.

So, in the strict sense of the word, only aircraft, birds, insects and bats fly, because winged flight depends on aerodynamic effects whereby lift is derived from the movement of air flowing round the outer surfaces of the body. In order to understand these effects, we must first consider the properties and behaviour of air.

Air and its movement

Both air and water are fluids, but air is a gas and water is a liquid. One of the differences between gases and liquids is that liquids are almost incompressible, while gases compress more easily. However, when considering low-speed flight, the matter of compressibility can for all practical purposes be ignored, as it does not assume any real importance until the object is approaching the speed of sound (about 760 mph). At this speed air becomes compressed or, to put it another way, changes density. The reason why this speed is significant is that sound is transmitted through the air in the form of waves which successively compress first one part of the air and then the next.

When an object is flying at a velocity less than the speed of sound, pressure waves travel out in front of the object and warn the air in front that the object is on its way. This enables the air to move from the path of the object and to pass to one side or the other. But when the object travels at above the speed of sound, things are very different because the warning wave is unable to travel fast enough to get in front of the object.

D1 A balloon demonstrating Archimedes' principle. Balloons will float upwards if their total weight, with passengers, is less than the weight of air they displace

Upthrust = weight of displaced air

Weight of balloon + weight of air

Now, the air instead of dividing and passing smoothly round the object strikes the object with a shock and becomes compressed. Needless to say, insects and birds do not have to concern themselves with supersonic flight and shockwaves in spite of the often quoted, but unaccountably inaccurate, observation that the deer bot fly travels at 800 mph. Nature has not found it necessary to travel at such speeds; supersonic flight by man is discussed in more detail in Chapter 5.

Movement of air provides the momentum that enables wings to overcome the force of gravity. We live at the bottom of an enormous ocean of air, which we tend not to notice until it starts to move; the faster it moves, the more we notice it. Although we have been using the movement of air for driving windmills and sailing ships for thousands of years, we do not appreciate its extraordinary power until it reaches very high speeds, as in the case of the terrible destruction wrought by a hurricane, or the remarkable aerodynamic forces necessary to support a 500-ton jumbo jet crammed with 400 passengers.

Winged flight can only function in our atmosphere, which is a comparatively shallow belt of air round the earth. As altitude increases, the atmosphere becomes progressively thinner; the maximum density occurs at ground level (D2). The air is held near the earth's surface by the force of gravity, or in other words by its own weight, which at sea level is about 0.077 lb. per cubic foot. Because of this weight, the air exerts a pressure on the earth of about 15 lb. per square inch at sea level; the further from the earth, the lower the pressure, so that at 20,000 feet it falls below 7 lb. per square inch. Since air is a fluid, its pressure is transmitted equally in all directions, upwards, sideways and downwards, including any space to which it can gain entry. But if, for some reason, the air pressure on one side of an object is reduced, either the object or the air will have a tendency to be sucked towards the direction of lower pressure. This is the principle behind winged flight.

Resistance, lift and drag

Moving air has very different properties from static air, for as soon as an object and air start moving in relation to each other, another force begins to exert its influence. It is so familiar that it is accepted without a second thought, and yet all winged flight, whether natural or mechanical, depends on it. The force in question is *resistance*, or to be more precise, the air's resistance to motion. It is curious that an understanding of its origin and nature came much later in man's history than of the distances and movements of the stars and planets. The effect of the air's resistance to an object can be felt in two ways: it can either appear as *drag* acting in the direction opposing that of the motion, or as *lift* acting perpendicularly to the direction of motion. When an aeroplane is flying at constant speed in straight and level flight, its thrust balances the drag, while its lift opposes the force of gravity by balancing the weight of the machine. In aerodynamics it makes no difference whether the body is at rest with the air flowing past it, or the body is moving through still air – it is the relative motion that counts.

The forces generated in a stream of air depend on several factors: air density,

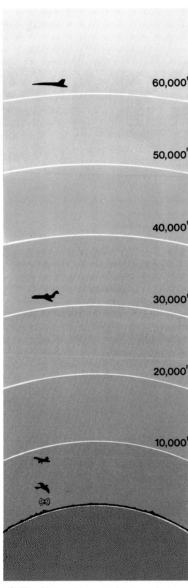

60,000'

50,000'

40,000'

30,000'

20,000'

10,000'

D2 The flight of insects, birds and aircraft can only take place in the comparatively shallow belt of air round the earth – the atmosphere

airspeed and the shape, the angle and the area of the surface meeting the air. The first factor is straightforward enough. If the density of the air is doubled, then twice the weight of air will flow over the surface, so doubling the forces generated. But with airspeed we find that the forces increase to the square of the velocity, that is, if the speed is doubled, then the aerodynamic forces increase four times. (This simple rule, however, does not apply at very high airspeeds of about the speed of sound.) If the area of the surface exposed to the airstream is doubled, then the aerodynamic forces produced will also be doubled. Finally, the relationship between the shape of the object and the angle at which it meets the airstream is much more complex and requires more detailed explanation.

Unless there is some obstruction, a stream of air will take the shortest route from one point to another of lower pressure. Any obstacle in the airstream will force the air to deviate from its normally straight path, producing a reaction on the obstacle in the form of resistance. The greater the deviation inflicted upon the air, the greater the resistance, so naturally the shape of the body in relation to the airflow has a significant effect. The best way to illustrate this is by drawing imaginary lines, *streamlines*, to indicate the direction of airflow at any given point. But the flow picture revealed by a pattern of streamlines is more than just a chart of flow direction, it also reveals the relative velocity of the airflow: where the streamlines are close together the velocity is high, and where they are separated the airflow moves more slowly.

A thin plate held with its edge towards the airstream offers minimum resistance because it causes minimum deviation in the path of the air, which follows a steady smooth route (D3). But the same plate held at right angles to the airflow will increase the air resistance, or drag, several hundred times (D4). In this position the air in front has to change its direction drastically, while behind the plate the airflow becomes broken up into eddies producing turbulence; the greater the turbulence, the greater the drag.

Drag can be dramatically reduced by *streamlining* the object, so that the turbulent spaces behind are filled in and the front areas are rounded or tapered (D5 & 6). By streamlining the flat plate its drag can be reduced by as much as 95 per cent. In fact, the filling in of the turbulent area behind has a greater effect on improving streamlining than altering the front. As drag increases with the square of the airspeed, it follows that the higher the speed required, the more streamlining is necessary. The ratio of the length to the breadth of a streamlined body is known as its *fineness ratio*; high fineness ratios improve streamlining efficiency, as is reflected in the extremely thin wings of supersonic aircraft. That part of the drag which is due to the shape of the object and which can be reduced by streamlining is known as *form drag*.

The nature of the body's surface also affects drag. It is easy to see that a rough surface will generate more friction as a result of the air flowing over it than a smooth one, although at low airspeeds this *skin friction*, as it is called, does not significantly contribute to the total drag. But with high-speed flight the effects of skin friction assume a far greater importance, so a smooth finish becomes an essential part of the design. If an aircraft is to meet with minimum air resistance, then the first consideration is to reduce form drag by designing the most suitable shape for the rear and front and by

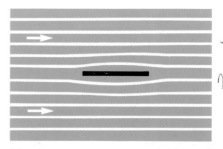

D3 Flat plate held edge-on to airstream

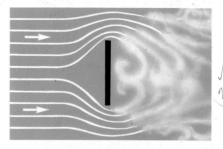

D4 Flat plate perpendicular to airstream

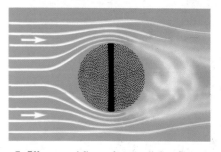

D5 Effect on airflow of streamlining flat plate

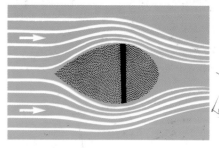

D6 Improved streamlining of flat plate

eliminating projections and sharp corners. After this, consideration can be given to reduce skin friction. A good definition of a streamlined body is one in which the form drag is less than the skin friction.

The angle at which a body meets the airstream, or *angle of attack*, is also a crucial factor. It is, however, important to realize that the angle of attack has nothing to do with the horizontal or the position of the earth; it is simply the angle at which the object meets the flow of air. Referring back to the thin plate, we know that there is minimum air resistance when the plate is held with its edge to the airstream, and maximum drag when held at a right angle; but what happens when the leading edge is inclined at a slight angle (D7)? The air pressure is now greater underneath the plate than on the top surface, producing a lifting force acting at right angles to the undisturbed airstream. But whenever lift is generated, drag is also created, and this acts in the opposite direction to the motion of the object. Whereas lift acts at right angles to the airstream, drag acts parallel to it. The net result is that the total force on the plate will be acting backwards as well as upwards (D8).

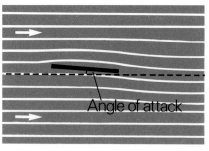

D7 Flat plate inclined to airstream showing angle of attack

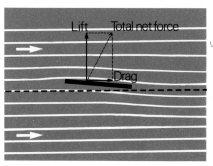

D8 Forces acting on flat plate

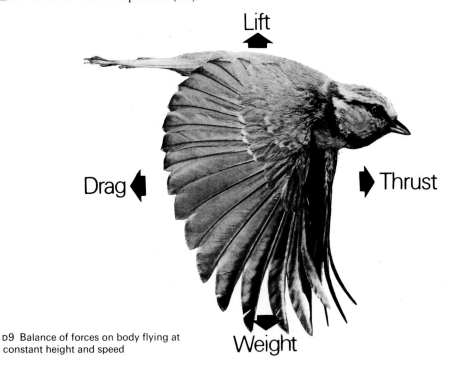

D9 Balance of forces on body flying at constant height and speed

Naturally, it is lift which makes heavier-than-air flight possible, while drag tends to prevent it. Both are really part of the same force, but as they have very different effects, it is important to distinguish between them. If an aircraft or bird is to fly at a constant height and speed, the lift acting on it must equal its weight, while the drag must be equal to the driving force or thrust. So the greater the lift, the greater the drag and the greater the thrust required to balance it (D9). An ideal aeroplane or bird would be all wing, free from projections and protruberances, such as fuselages, tails, engines, heads, wheels, legs or any other extra parts which produce drag without directly contributing towards lift. Drag in this form is appropriately called *parasitic*, and in aircraft everything is done to keep it to a minimum. But the drag which is actively responsible for producing lift, the *induced drag*, cannot be eliminated, even in an ideal aircraft, as it is a necessary adjunct of lift. More will be said about this later.

Wings and aerofoils

The whole principle of wings, whether they be of bird, insect or aeroplane, is that they are designed in such a way as to push the air downwards, and in so doing gain an upward reaction. It is rather like walking up a sandy slope, every step upwards pushes the sand downwards. Both examples obey one of Newton's laws that, 'To every action there is an equal and opposite reaction.' The downward flow of air created by wings is called *downwash*. The more air deflected downwards by a wing, the greater the lift; but on the other hand the more air disturbance produced, the greater the drag.

Although a flat shaped object can function as a simple wing, as demonstrated by the thin balsa-wood wings of toy gliders, a wing's lifting and general aerodynamic characteristics can be dramatically improved if it is shaped and streamlined in a certain way. Wings which are designed and angled to the airstream in such a way that the maximum downwash and lift is obtained with the minimum of turbulence and drag are known as *aerofoils* (D10). The gradual curvature of the aerofoil entices the air to flow smoothly over the upper surface in a downward direction without breaking away and forming eddies. This is how the aerofoil scores over the flat or cambered plane (D11).

The peculiar properties of aerofoils rely on a natural law concerning the relationship between the speed at which a fluid moves and the pressure it creates. Formulated by the Swiss mathematician, Daniel Bernouilli (1700–82) over two hundred years ago, and known as the Bernouilli principle, this law affirmed that as the velocity of a fluid increased, the less pressure it exerted. The principle can be demonstrated quite simply by holding two pieces of angled paper an inch or two apart, and by blowing between

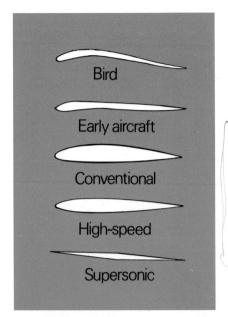

D10 Aerofoil sections and the parts of a wing

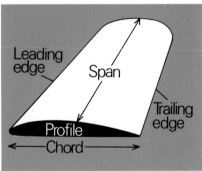

D11 Ratios of lift to drag on flat plate, cambered plane and aerofoil

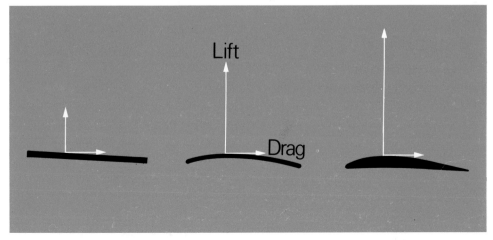

P7 & 8 Bernouilli's principle demonstrated by blowing between two sheets of paper

P9 Instead of pushing the spoon away, the jet of water sucks the spoon inwards

them. What happens is that instead of opening up the two sheets close together (D7 & 8). Because the air is incompressible, it has to accelerate to pass through the restriction, thereby reducing the pressure inside, and so the two sheets are forced together by the relatively higher pressure on the outside surfaces. Another experiment demonstrating Bernouilli's principle can be done by holding a tablespoon downwards between finger and thumb, and directing a jet of tap water onto its convex surface; instead of pushing the spoon away the water will hug the curvature, pulling it into the jet (P9).

Both these experiments illustrate the drop in pressure which occurs when the speed of air or water increases. The shape of a spoon shows a marked similarity to the wing of a bird or aeroplane, and in fact all three behave as aerofoils. When a stream of air passes round an aerofoil, the air flowing over its more convex upper surface has a greater velocity and, following Bernouilli's principle, a lower pressure than that on the under surface; the difference in pressure between the two creates the lift (D12). Unlike a flat plate, an aerofoil with a humped upper surface will produce lift even when the angle of attack is 0°. Further lift is gained by increasing the angle of attack so that the air meets the under surface at a steeper angle (D13).

The lift derived by a wing is the sum total of all the pressures acting upon its surface. It is by no means equally distributed, as normally about two-thirds is due to the decrease in pressure on top, and one-third due to the increase in pressure underneath (D14). However, for the sake of convenience, the total aerodynamic force can be considered to act from a point known as the *centre of pressure*, although its position alters with the angle of attack (D15).

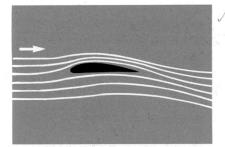

D12 Flow of air around aerofoil section

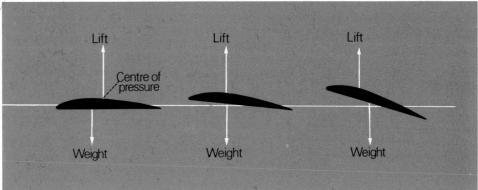

D15 Movement of centre of pressure with angle of attack

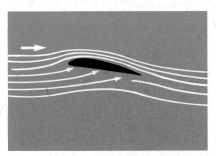

D13 Effect of increasing the angle of attack

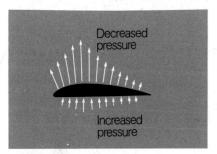

D14 Pressure distribution over aerofoil surface

For any given wing section, lift, drag and centre of pressure are largely controlled by the angle of attack. As the angle of attack increases, so does lift, while the centre of pressure moves forwards. Unfortunately, drag increases at the same time, slowly at first, but at a certain point it starts to increase more rapidly than lift. Any particular wing, therefore, has an optimum angle of attack, providing the best ratio of lift to drag; most aircraft are designed to cruise at this angle because in this way they consume least fuel. Typical aircraft aerofoils have their most efficient lift-to-drag ratio at about $3\frac{1}{2}$ to $4°$.

As the angle of attack is increased, the gain in lift cannot be indefinite because a stage is reached when the air ceases to flow smoothly over the wing's upper surface. Although the under surface may still be providing lift, the air on the upper surface is no longer able to flow smoothly round the hump, but breaks away into turbulent eddies. At this point the wing loses its lifting properties and is said to *stall*. The angle at which stalling occurs is called the *stalling angle*: in aircraft this is at about 15°, which is much higher than with a flat plate, in some insects it may be as high as 60° (D16). In an aircraft the suddenness with which this comes about can be dangerous, particularly as the pilot has little direct control over the angle of attack. It should be remembered that the angle of attack depends on the relative airflow, thus a stall can occur at any altitude. For instance, if an aeroplane is flown too slowly, it will sink, and so meet a relative airflow from underneath causing the stalling angle to be reached, even though the angle of the aeroplane may be horizontal to the ground. For this reason, it is more usual to talk about *stalling speeds* than stalling angles. It is important not to confuse stalling with sinking; any aircraft or bird will sink when the total upwards force is less than its weight, but stalling takes place when the angle of the attack or the air speed is such that the air suddenly separates from the upper surface of the wings and becomes turbulent. The more practical aspects of stalling are discussed in later chapters, together with the various devices used by birds and aircraft for reducing their stalling speeds.

As well as providing lift, the difference in pressure between the lower and upper wing surfaces is responsible for a related effect known as *wing-tip vortex*. Explained simply, wing-tip vortex is due to the tendency of air, or any fluid, to flow from high to low pressure. As there is no barrier at the wing-tips separating the high from the low pressure areas, the air leaks round from underneath the wing to the top surface (D17). This wing-tip flow causes the air on the top surface to be deflected slightly inwards, and that on the bottom surface to flow outwards (D18). As a result, the streams meeting at the wing's trailing edges cross one another to form a series of small trailing vortices, which then join up into one large vortex at each wing-tip (D19). The energy which goes into the formation of these vortices appears as the induced drag, which has already been mentioned. Wherever there is lift, there must also be induced drag, as it is caused by the wings which are active in generating lift. Unlike other types of drag, induced drag does not increase with the square of the airspeed; on the contrary, it is at its greatest when maximum lift is being achieved at the lowest airspeed, that is just before stalling. This is because as the airspeed is reduced, the angle of attack has to be increased to maintain lift, which induces a greater difference in pressure between the upper and lower surfaces of the wing. Consequently, there are more violent wing-tip vortices and, therefore, more induced drag.

One way of keeping induced drag low is to extend the wings and taper the tips, so that

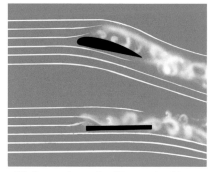

D16 Comparison of stalling angle of flat plate with aerofoil

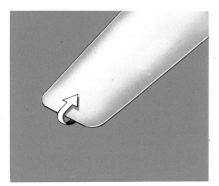

D17 Air leaking from underneath a wing to the top surface

D18 (*left*) The cause of trailing vortices – the deflection of the airflow above and below the surface of a wing

D19 (*below*) Trailing vortices

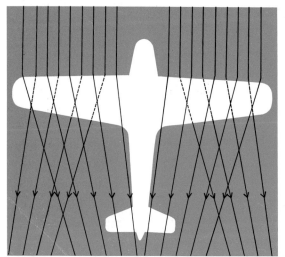

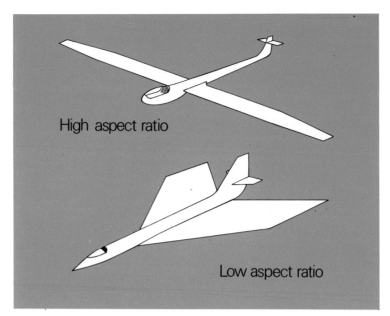

the tips make up a relatively small proportion of the wing. Aircraft and birds with long narrow wings have what is called a *high aspect ratio,* which is the ratio of span to chord (that is, length to breadth), while those with short stubby wings have a *low aspect ratio* (D20). The main advantage of high aspect ratios is that by cutting down the induced drag they provide maximum lift with minimum consumption of power. That is why gulls, gliders and aeroplanes designed for long-distance flying have high aspect ratios – long narrow wings are simply more efficient, particularly at low speeds. But like everything else, high aspect ratios have their snags: as well as being more difficult to manoeuvre both in the air and on the ground, structurally they tend to be heavier, so that a point is reached when the benefits of increased lift are cancelled out by the extra weight. This is one of the main reasons why man-powered aircraft have not been more successful. To minimize the induced drag, which at low airspeeds consumes a high proportion of the power, it is essential to have wings with high aspect ratios, but the extra weight they carry imposes a practical limit on the wing-span.

Any means of reducing drag without sacrificing lift or, conversely, increasing lift without increasing drag, must obviously improve the lift-to-drag ratio. An alternative way of minimizing induced drag at low speeds is by having a large wing area for supporting the weight of any given body or, in aeronautical language, *low wing-loading.* This simply means that as the surface area of the wing is larger, each square foot has less weight to support. The chief disadvantage of low wing-loading is that as the wings are relatively large, they create more surface friction. As skin friction increases with the square of the airspeed, birds and aircraft that are efficient at high speeds have small wings and, hence, *high wing-loading.*

Stability

Flight cannot be achieved simply by attaching a power unit to a wing and taking off, as it is obvious that such a contrivance would lack *stability* and be hopelessly uncontrollable. Any vehicle, whether it be a car, ship or aeroplane, requires some degree of stability. A two-wheeled handcart is unable to remain balanced unless stabilized by a driver through a shaft (D21). Likewise, a wing would be equally unstable unless balanced in

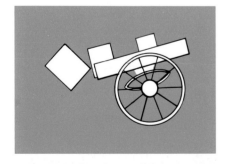

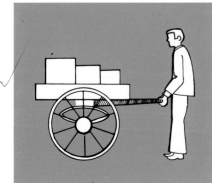

some way. Imagine a wing flying along with its lift and weight perfectly in balance; the equilibrium would be upset by the slightest air disturbance. If, for instance, the front of the wing is lifted up, the angle of attack increases (D22). Remembering that the centre of pressure shifts forward with increasing angle of attack, so the two lines of force become out of line with each other. This produces an even greater angle of attack, resulting in a further forward movement of the centre of pressure, inducing the wing to start revolving uncontrollably round its long axis. A practical demonstration of this can be made by dropping a postcard from a height.

The matter can be rectified quite simply by the addition of a fuselage, which acts as a lever in the same way as the shaft of the handcart, while in place of the controlling hand a *tailplane*, or *stabilizer*, is added. The function of the tailplane is to create a force in the appropriate direction to compensate for any out-of-balance effects, providing what is known as *longitudinal stability*. Normally, the tailplane provides no lift, by virtue of its symmetrical aerofoil section and zero angle of attack, but as soon as a disturbance causes a nose-up attitude, the tail drops and a positive angle of attack is produced. In this way, lift is generated, pulling the tailplane upwards, and the wings are brought back to their original position (D23).

As well as affording longitudinal stability, measures also have to be taken to prevent aeroplanes from wandering from one side or to the other. *Directional stability*, as it is called, is achieved by including a vertical surface, the *fin*, above the tailplane. The fin functions in the same way as the tailplane, but corrects in a directional sense. It should also be borne in mind that the sides of the fuselage have a larger area behind the centre of gravity than in front, so that as soon as the machine 'weathercocks' to one side, the pressure differential assists the fin in rotating the machine back to its original path (D24).

Finally, some method has to be found to maintain *lateral stability* so that the aeroplane remains on even keel and is prevented from rolling from one side to the other. If the machine begins to roll, forces come into play to oppose the roll, because the wing which is moving downwards effectively strikes the air at a greater angle of attack than the upward moving wing. As each wing is now subjected to a different amount of lift, there will be a tendency for level flight to be restored automatically. However, once the aeroplane has tilted into a roll, its weight and lift become laterally out of line with each

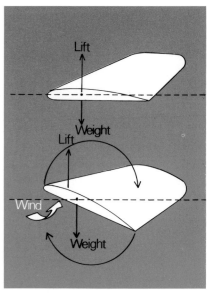

D22 Instability of a simple wing

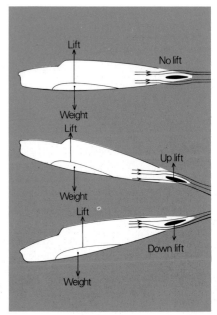

D23 Stabilizing function of tailplane

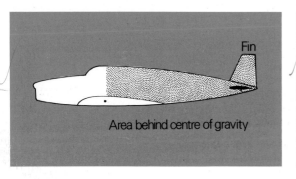

Fin

Area behind centre of gravity

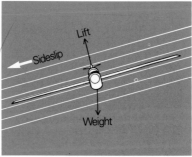

Lift

Sideslip

Weight

D24 (*left*) Directional stability provided by fin and side of fuselage

D25 (*right*) An aircraft in a sideslip

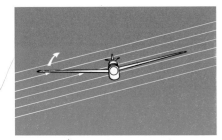

D26 Use of dihedral for lateral stability

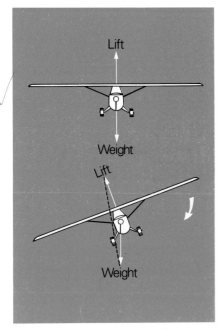

D27 In a high-wing monoplane the centre of gravity is kept low so that the machine behaves like a pendulum

other, resulting in the plane sideslipping towards its lower wing (D25). Now, if the two wings on each side are parallel, the aircraft will have a tendency to carry on sideslipping because the angle of attack on each span is the same. But by incorporating a *dihedral*, so that the wings are arranged in a flat 'v', the angle of attack on each wing of a sideslipping aircraft is different (D26). It is easy to see that the effect of the lateral airflow will be to raise the lower wing so that the aircraft assumes its former level position. An alternative method of establishing lateral stability is achieved in a high-wing monoplane where the wings are positioned on the top of the fuselage (D27). The centre of gravity is kept low so that the machine behaves like a pendulum, counteracting the lateral tilt of the wings.

All round inherent stability, which is the combination of all the stabilities mentioned, makes the pilot's task very much easier, particularly in turbulent conditions when the aeroplane is being tossed around in the sky. But as we shall see in a later chapter, there was a time in the early days of flying when unstable aeroplanes were preferred to stable ones, which meant that the pilot was constantly fighting the controls to keep his machine in the air. Nevertheless, stability should not be excessive, particularly with aircraft which need to be highly manoeuvrable, because they would then have a tendency to resist any change in attitude or direction of flight. Insects and birds, on the other hand, depend for their survival on being able to manoeuvre rapidly and do not require the inherent stability of aircraft because their flight is continually monitored and adjusted by reflex action.

Boundary layer

Because of its internal friction, air, or any fluid, has a resistance to flow smoothly – in other words, it has viscosity. For example, treacle is more viscous than water and both exhibit different degrees of resistance to rate of change in shape. This is due to the tendency of one layer to 'stick' to the adjacent layer, thus resisting relative movement between the two. Air is plainly much less viscous than treacle or water but, nevertheless, what viscosity it has is enough to cause skin friction; ultimately it is responsible for causing all turbulence, drag and even lift itself.

One of the greatest advances in the history of aerodynamics was made early in the twentieth century by a German mathematician, Ludwig Prandtl (1875–1953). His discovery of the *boundary layer* introduced a new concept about the motion of fluids, which has subsequently led to significant improvements in the design and performance of aircraft. We know that if drag is to be kept low, then it is essential that the air is induced to flow smoothly over all the surfaces with which it is in contact. But flow of air immediately adjacent to a surface has peculiar properties of its own, in that it never actually flows over the surface at all.

The flow of air within the boundary layer is similar to the movement of the leaves of a book lying flat on a table; when the upper corner is pushed at right angles to the spine, the pages slide over one another, the speed of the movement of the pages decreasing from the top downwards. The volume of the book is said to be *sheared*, while the force that causes it is termed a *shearing stress*. The concept of the boundary layer assumes that, however smooth a surface may be, the molecules of air which are in actual contact with the surface remain motionless and do not move over it. Some distance above the surface the air moves smoothly and at full speed, but sandwiched between this main flow and the stationary film is the boundary layer, where the velocity of the air molecules increases outwards from the surface (D28).

This velocity gradient is confined to an extremely shallow layer of air which is normally about one hundredth of an inch thick, and no matter how rapidly the object is moving, whether it is a bullet or a supersonic fighter, the relative velocity of the air at its surface is exactly zero. As the airflow above the boundary layer is moving at full speed, it is not difficult to imagine the magnitude of the shearing stresses which must be created within. If the successive layers of molecules within the boundary layer slide smoothly over each other, then the flow is said to be *laminar*. But on the other hand, if the various layers are dragged along on top of one another forming rolling eddies and vortices, than the flow is turbulent, resulting in increased skin friction and wasted energy. When the boundary layer breaks away from the surface, it gives rise to increased drag or stalling; if it can be controlled by keeping the flow smooth and close to the surface, drag can be kept to the minimum and the stalling point delayed. There are a number of mechanisms employed both by flying creatures and aeroplanes which increase lift and reduce stalling speed by maintaining a laminar flow in the boundary layer. These are described in later chapters.

When air flows over a wing, it usually does so in a laminar manner up to the thickest point, but after this the flow tends to become turbulent within the boundary layer. Laminar-flow wings on aircraft are designed to preserve a smooth flow within the boundary layer as far back as possible (D29). In this way, the transition from smooth to turbulent flow can be considerably delayed, reducing drag due to skin friction by sometimes as much as 50 per cent. It is likely that the behaviour of the boundary layer on the wings of insects plays a crucial role in the micro-aerodynamics of insect flight.

Non-steady airflow

Aircraft and the majority of flying creatures fly in what might be called a standard way, using well-understood aerodynamic principles. However, certain creatures, particularly insects and small birds, are able to perform in a manner that cannot be explained in simple aerodynamic terms. For instance, until recently nobody had been able to explain how hoverflies hover, how dragonflies fly backwards or hover or why bumblebees are able to fly at all, as calculations were based on conventional textbook aerodynamics. But recent research at Cambridge University by the late Professor Torkel Weis-Fogh has revealed some fascinating new evidence which has helped to explain how the insect world has been apparently violating some of these established principles. He has discovered that insects make use of non-steady airflow to generate lift. The details of the actual mechanisms used are explained in Chapter 2, so here we shall only discuss the broad outlines.

We know that aircraft depend on aerofoils which move through the air steadily. Any non-steady airflows round aircraft wings have to be minimized, since they reduce the

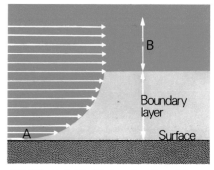

D28 Effect of skin friction showing airflow decreasing speed towards surface. At A the air is static and at B it is moving at full speed

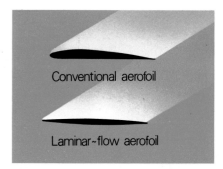

D29 A comparison between a conventional aerofoil and a laminar-flow aerofoil. In the laminar-flow aerofoil the maximum thickness of the profile is around the centre, rather than towards the front

efficiency of flight and may even produce dangerous vibrations. In contrast, insects and birds depend on flapping wings which produce varying degrees of non-steady airflow. Thus, non-steady aerodynamics is an inherent feature of natural flapping flight, but its importance varies with the size and airspeed of the animal.

Generally, the larger and faster the animal and the lower the frequency and amplitude of its wing-beats, the less significant are the effects of non-steady airflow. It can be assumed, therefore, that when insects and birds are in fast forward flight, the flow of air around their wings is almost steady and their performance can be explained in terms of ordinary aerodynamics; in other words, the airflow can be understood to work as a sequence of steady-flow situations.

The flight of the smaller and slower-moving creatures, however, is different. Their wing movements are far more extensive and have a faster frequency, resulting in aerodynamic forces which vary considerably in magnitude and which oscillate widely in the three dimensions of space. So, the airflow becomes non-steady, the fluctuations increasing in intensity and significance as the airspeed diminishes, reaching a maximum in hovering flight. Hovering is not confined to hummingbirds, kestrels, hoverflies or dragonflies. Most small birds can hover for short periods, particularly when courting or approaching their nests. Even certain small bats are capable of hovering, and among the insects the ability to hover is the rule rather than an exception. Although there is more than one type of hovering, it is usually a strenuous form of flight and makes peculiar aerodynamic demands upon the creature in question.

The conditions under which oscillating wings function vary according to the size of the animal. Aerodynamicists describe the relationship in terms of *Reynolds numbers*, but as its concept is difficult to understand in non-mathematical terms, it will not be explained here, beyond saying that it is the ratio of inertia to viscous forces. The value of Reynolds numbers for flapping wings various considerably from about 15,000 in hummingbirds, to well under 20 in some tiny parasitic wasps, and for large aircraft in high-speed flight it can exceed 20,000,000. The range of these numbers is important because, although normal aerofoil action functions well at high and intermediate values, it deteriorates dramatically toward the lower end of the scale, when the viscous shearing forces and drag are large compared with the inertial forces and lift. In effect, this means that the smaller the animal (and the lower the Reynolds number), the larger the relative drag. So how is it that animals hover by means of flapping wings? And how is it that very small insects remain airborne at all? Before answering these questions, we must look at the nature of lift from another angle.

If we could see a stream of air as it flowed past an aerofoil, it would appear as if the air were actually circulating round it. Consider, for example, a horizontal revolving cylinder. If the cylinder is placed in an ideal fluid of zero viscosity, its rotation would not affect the fluid in any way as there would be no friction or boundary layer produced. But when the cylinder is immersed in a real fluid which has viscosity, such as air, and spun round its axis in an anti-clockwise direction, the surrounding air is set in motion and rotates with the cylinder; this is called a *bound vortex* (D30).

If a horizontal wind were now to blow from right to left, the combination of the

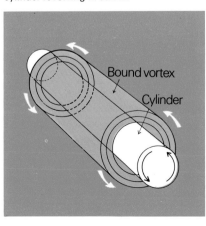

D30 Formation of bound vortex around cylinder revolving in still air

anti-clockwise rotating air of the bound vortex and the horizontal airflow would cause an increase in the speed of the air above (compression of streamlines) and a decrease in the speed below. The net result would be a decrease in pressure above and an increase below, due to the usual Bernouilli effect, together with an upwash in front and a downwash behind; in effect, the spinning cylinder would be subjected to lift in the same was as an aerofoil (D31). This phenomenon is known as the *Magnus effect*, and a familiar example of a similar kind of action can be seen when a sliced golf ball is sent sailing into the rough (D32). As the ball is driven forward, it is set spinning by an oblique blow of the club, and the effect of viscosity induces the air to stick to the ball's surface and rotate with it. Where the rotary motion of the bound vortex and the airflow are in the same direction on the righthand side of the ball in the diagram, the combined streams move rapidly; and where they clash head-on on the left, the velocity is reduced. The resulting pressure differential induces the ball to deviate from its path. It is important to understand that without the rotational flow of air (bound vortex), and its superimposition upon the airstream as the ball flies through the air, there would be no lifting force at all.

With this in mind, we can now study aerofoil action in a new light. We can think of it as a device which acts like the revolving cylinder for creating and maintaining circulation in the form of a bound vortex. If we were able to move with the undisturbed airstream and

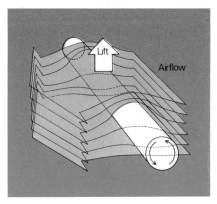

D31 Lift generated by cylinder revolving in horizontally moving airflow

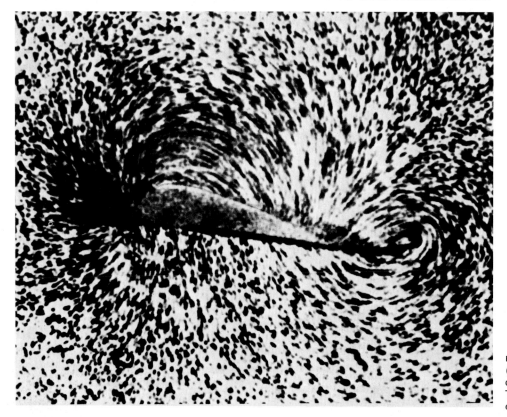

P10 Circulation of liquid round an aerofoil. (The photograph was taken with the camera moving with the flow of liquid.) The starting vortex is visible at the trailing edge

D32 Sliced golf ball, with an anti-clockwise spin, being subjected to a lifting force to the right

watch the airflow moving over a wing, it would seem as if the air was circulating (P10). The air in front moves upwards and over the top of the aerofoil flowing faster than the main airstream, while the air underneath is flowing slower than the main airstream; so the air, relatively speaking, appears to move in a circle. The idea of relative airflow round an aerofoil is called *circulation*, but the particles of air do not actually circulate round the wing. The circulation, as we know, in fact takes place outside the thin boundary layer, although it is the boundary layer that initiates this flow.

Provided circulation is established, a wing will experience lift as soon as it is exposed to a wind, but a knowledge of the way in which the circulation is set up in the first place is essential for understanding insect flight. What happens when a wing starts to move through the air from rest? By holding a piece of inclined cardboard in smoke and moving it from rest, an eddy will be seen to be shed from its trailing edge (D33). This is called a *starting vortex*, and it is created because of the large viscous shearing forces close to the trailing edge of the wing; it is generated every time a wing starts its

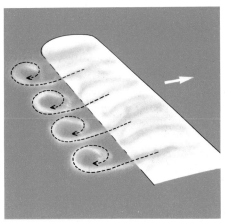

D33 Starting vortex on the trailing edge of a wing

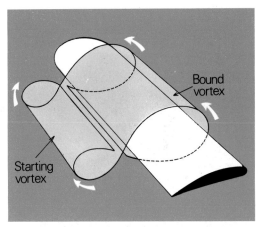

D34 Starting vortex and the production of its counter vortex, the bound vortex

movement. One of the rules of aerodynamics is that a vortex cannot be created without the production of a counter-vortex of equal strength circulating in the opposite direction (D34). In this case, the counter-vortex is in fact the bound vortex which is responsible for the circulation and production of lift, and it owes its continuing existence to the shearing forces over the surfaces of the wings. At the start of the wing movement we now have to consider two opposing currents of circulating air, the starting and the bound vortex. Since they interact destructively in inverse proportion to their distance apart, the net lift around the aerofoil at the start of the movement is very small. But once the starting vortex has been left well behind (after the wing has moved about three chord lengths away from its starting point), then the smooth characteristics and lift of steady airflow are established. The interaction which delays the creation of lift is called the *Wagner effect* and constitutes an unavoidable non-steady phase in the action of normal aerofoils.

Let us now consider the effect on flapping wings. During hovering in particular, the repeated stopping and starting at the end of each wing stroke would appear to hamper, rather than assist, the production of lift. This becomes particularly acute with small insects at low Reynolds numbers, when the creation of circulation becomes difficult and, furthermore, the vortices tend to die out rapidly. Another apparent disadvantage of the flapping wing is that when it comes to rest at the end of each stroke and the lift is reduced, the bound vortex has to be shed, becoming a 'free' vortex, which in hovering flight will interact with the new starting and bound vortices that initiate the return stroke. Calculations made of the minimum lift coefficients necessary to sustain flight at low Reynolds numbers show them to be far in excess of what is possible in a steady-flow situation.

Clearly then, standard aerodynamics has failed to furnish the answer as to how insects and birds overcome these apparently insuperable handicaps. But thanks to the published researches of Professor Weis-Fogh and the evidence of high-speed photography of hovering flight, the mystery today is nearer solution. It is now possible to conclude that insects and birds turn non-steady airflow to their advantage by employing mechanisms which can swiftly establish air circulation round their wings, entailing the rapid formation and shedding of vortices. By so doing, they are able to generate far more lift than would be possible in steady airflow. This discovery has cast an entirely new light on the problems which have long perplexed observers of bird and insect flight, and is discussed in greater detail in the subsequent chapters.

2
The first to fly

THE SIGHT AND SOUND OF WINGS are so much part of our everyday experience that we tend to take for granted anything in the air from dancing gnats to supersonic fighters. Indeed, it is difficult to imagine what it would be like to have no wings about us, to look up into a sky empty of insects, birds and aeroplanes; yet up to a time unimaginably distant, the sky was a lifeless silent void. Then, from the dense undergrowth of steamy swamps and from the edges of lagoons and lapping seas anonymous insects began their first tentative flights – flights even more momentous and crucial for the destiny of life on this planet than those at Kitty Hawk and Cape Kennedy some 350 million years later.

At the time of this evolutionary breakthrough, the earth had already been supporting life in various forms for more than 1,000 million years. Flightless insects had been sharing fern-pond and mud-flat with scuttling millipedes, spiders and scorpions for perhaps 25 million years; but with the arrival of the flapping wing, insects gained an advantage over all other creatures which was to lead to a species and population explosion unmatched in size and duration. Unquestionably, the aerial performance of insects is quite spectacular by any standards, yet far less is known about their flight than that of birds or aeroplanes. Moreover, until recently even the fundamental laws of aerodynamics appeared to break down when insect flight was considered.

Evolution of wings

The evolution of these first wings took place at about the same time as the first amphibians abandoned the water for a life on land. Insects and other established earthbound creatures probably were a convenient source of food for the newcomers, and so it is quite possible that the amphibians provided the insect with the evolutionary challenge which led to the development of wings. There are no fossil records of flying insects until the late Carboniferous period (some 270 million years ago); the details of their wing evolution before this time cannot be accurately traced, partly because early insects were particularly fragile creatures and disintegrated before becoming fossilized.

Although there are other theories, insect wings are generally thought to have evolved from lateral extensions at the top of the thorax wall, appearing first as shallow flaps which may have served as stabilizers as they jumped out of harm's way (D35). Important supporting evidence comes from one of the earliest insect fossils, which actually exhibits two small lobes on the first segment of its thorax, in addition to the two pairs of

D35 Hypothetical early gliding insect

fully developed wings on the other two segments. It would seem as though all three pairs started to evolve together, but the front ones never got very far in their development, the idea having been eliminated through the processes of natural selection at an early stage. After all, six asynchronous flapping wings might have presented aerodynamic problems of a complex order.

As the process of evolution continued, the stabilizing flaps gradually became larger, developing into flat planes which helped the insect glide to the ground. It is possible that these early fliers may have been strong jumpers like modern grasshoppers or creatures like cockroaches that leapt from the sides of tree trunks to plane away to safety. But having reached the gliding stage, their wings had to become movable before they could learn to flap them. Perhaps the first muscles developed for folding the wings back as the insect crawled through tangled vegetation, but, whatever their origin, they formed the basis for the development of the elegant and efficient mechanisms of today's insects.

According to the earliest fossil records, by the late Carboniferous period there existed a variety of fully flying insects. Most were primitive cockroaches, mayflies and gigantic dragonflies, some with a wing-span approaching three feet; but the most antiquated one of all, both in time and appearance, was the ancient net wing (D36). This clumsy insect was from the order *Palaeodictyoptera*, a group which disappeared during Permian times (which began 270 million years ago) when several orders of modern insects first appeared.

The Permian period, which heralded the spread and diversification of the reptiles, was a prolific time for insects, with grasshoppers, leafhoppers and several other familiar insects filling the air. Pollinating insects did not evolve until the flowering plants started to grow about 130 million years ago, when butterflies, moths, bees and flies began to flourish and rapidly to assume an ever-increasing variety of forms. After thousands of years of search and discovery for new habitats and food supplies, over which time complex life histories have evolved, the more highly specialized insects of today behave in a way which can only be described as awesome. Some minute wasps, for example, can only exist as a parasite in the body of a parasite which is in the body of another parasite. Certain orders of moths have sensory apparatus for detecting the ultrasonic squeaks of bats which 'see' in the dark by using these sounds as a sonic radar, so that by making abrupt changes to their flight-paths they can avoid capture. Even more remarkable is the fact that at least one group of inedible moths has developed the ability to produce ultrasonic clicks of its own; these are emitted in response to those of the bat and act as an early warning signal, saving not only the victim, but also the hunter, from a distasteful mouthful.

Why the insects have succeeded

Of all creatures in the animal kingdom, insects are the most successful. As well as outnumbering by five times all other known species of animals put together, they have invaded, and dispersed, to every corner of the earth, their numbers multiplying as the

D36 The ancient nett wing (*Palaeodictyoptera*)

P11 Dragonflies (*Odonata*) are among the most primitive of modern insects – some prehistoric species of the Carboniferous period had a wing-span of nearly 36 ins

climate becomes hotter. Although the insects' success can be attributed to a number of factors, their conquest of the air is the most important. With the exception of birds and bats, the ability to fly sets them apart from all other creatures. If conditions become unfavourable in one place, if they are short of food or threatened by enemies, they simply take to their wings and find a more hospitable habitat elsewhere.

There are also at least three other major factors which have contributed to their success: their small size, metamorphosis and external skeleton. A crumb too tiny to be noticed by larger animals can provide a feast for an insect, a drop of dew quench its thirst and a fallen leaf protect it from the midday sun. Many insects are able to occupy minute niches in the environment, such as the inside of seeds or the tissue between the upper and lower surfaces of a leaf, which are obviously quite unsuitable for larger animals.

Second, the majority of insects benefit from having what is known as a *complete metamorphosis*, when their lives are divided into two active stages, each one feeding on entirely different foods. In this way, an area can support far greater numbers than it could if the insect consumed one type of food throughout its life. Take as an example the leaf-eating caterpillar and the nectar-sipping butterfly: growth only takes place in the caterpillar, or larval stage, where the creature simply becomes a super-efficient organism for gathering food. After the dormant intermediate pupal stage, the chrysalis, the insect becomes transformed into a butterfly attuned to a life of reproduction. The mature insect is now complete with wings and reproductive organs, allowing it to devote the remainder of its life to finding a mate and laying eggs in an area favourable for the growth of its offspring.

The insects' third asset is the possession of a hard, but lightweight, outer skeleton, the *exoskeleton* or *cuticle*, which serves as protective armour and an attachment for muscles. It is largely composed of a horny substance called *chitin* which, as well as providing strength, makes for an efficient two-way waterproofing system, not only keeping water out, but above all protecting the insect from dehydration. To improve the efficiency of the waterproofing the outside of the cuticle is covered by a thin layer of wax. The problem of drying out is particularly severe in small animals, such as insects, as the smaller the body is, the larger its surface area compared with the volume. Consequently, the loss of water by evaporation becomes relatively very much higher as the animal becomes smaller.

Together with flight, these three factors combine to give insects a degree of adaptability not found in any other form of animal life, allowing them to take advantage of almost any conditions on earth. They have been found flying at over 15,000 feet, and in one species, a midge from West Africa, the larva is not only able to survive being frozen in liquid air at 190 C, but also being boiled in water for short periods. Insects have too exploited practically all possible food supplies, from green plants and roof timbers to paint brushes and pools of crude petroleum.

Form and function

Insects have a greater variety of form and function than all other animals, but their basic structure is uniform. Compared with birds, insects are incongruous creatures, as apart from wearing their skeletons on the exterior, their bodies are made up of a series of segments connected to one another by flexible joints. Adult insects are divided into three main sections: the head, the thorax and the abdomen, each with their own functions (D37). The head, carrying the mouthparts and principle sensory organs, such as the antennae and eyes, is mainly responsible for orientation and feeding. The thorax, with its massive musculature, is almost exclusively concerned with locomotion; attached to its three segments are three pairs of legs, and usually one or two pairs of wings. The abdomen contains the digestive, excretory and sexual organs, together with the insect's long tubular heart and most of the breathing mechanism.

D37 Structure of a typical insect

These basic characteristics distinguish insects from all their arthropod relatives, such as spiders, crabs and centepedes, which all have more than three pairs of legs (spiders have four) and a body divided into either two main divisions, or more than three. However, as in other arthropods, the blood system of insects has few vessels; most of the blood is contained in large cavities throughout the body where it slowly circulates, freely bathing the various organs. It has no red corpuscles and plays no part in carrying oxygen around, its main functions being to clear the body of bacteria and particles from cell breakdown and to transport fuel, hormones and nutrients to the tissues.

P12 Like other primitive flying insects, the wings of scorpion flies (*Panorpa*) are powered by direct flight muscles – the independent movement of the two pairs is clearly shown in this photograph and the hindwings are folded so far forwards that they are nearly touching each other

Since the blood does not absorb oxygen, insects do not need lungs like mammals and birds, and so have evolved a totally different method of respiration. Air seeps into the insect's body through small openings in the cuticle, called *spiracles*, and is distributed to all parts of the body through a maze of microscopic air tubes, known as *tracheae*, which ramify throughout the tissues. The oxygen is not taken in through the walls of the tracheae, but is absorbed in fluid at the ends of their finest branches. To ensure a sufficient and accessible supply of oxygen, it is probable that these minute branches actually penetrate nearly every cell of the muscles and other tissues.

It used to be thought that air could only be conveyed through the tracheae by the simple process of diffusion, and as diffusion only worked over short distances, it accounted for the limit in the size of insects; now it is understood that the majority of insects do not rely solely on this simple form of respiration, but have a ventilating system which forcibly drives air through their bodies, in a similar way to our breathing. Some of the tracheae open out to form large thin walled air sacs which collapse and expand as the result of a slow rhythmic muscular action by the abdominal segments. Such movements are easy to see in a resting hoverfly or wasp. Furthermore, it has been demonstrated that in certain insects the main flow of air is unidirectional: when the abdomen expands, air enters the body through one set of spiracles and when the abdomen contracts, it passes out through another, the rate of ventilation increasing as the insect's activity becomes more vigorous. By far the most strenuous exercise is flight, and here the considerable pulsating distortions of the thorax caused by the flight muscles act as a pump. In this way, the thoracic air sacs are made to function like bellows, driving the air in and out of the tracheal system and thereby keeping the rate of oxygen supply to the flight muscles in pace with the insect's flying activity.

Controlling temperature for flight

Before an insect is able to take to its wings, its body temperature has to be sufficiently high; in the majority of insects the flight muscles are unable to develop full power until they are at least 25°C. Birds have automatic mechanisms for maintaining body temperature and so are not limited by the temperature of their surroundings, but with insects it can, and normally does, rise and fall according to that of the air. Physical activity is related to these conditions, so that insects become slow in movement, and even torpid, when cold. Thus, the majority of insects are only active when ambient conditions are hot enough, often relying on the sun's radiant heat to keep their bodies at a suitably high temperature for flight, mysteriously disappearing from sight as soon as the sun disappears behind a black cloud. When the sun reappears, it is the smaller insects which take to their wings first, for they warm up faster than the larger species due to their higher surface-to-volume relationship. Many insects can to some extent regulate their temperature by controlling the area of the body exposed to the sun's rays: grasshoppers, for example, face the sun during the heat of the day, but turn broadside on when the sun is low.

Large-bodied insects, such as silk moths, hawk moths and bumblebees, have found

P13 The spurge hawk moth (*Celerio euphorbiae*), like other members of its family, is beautifully streamlined and is one of the fastest insects on the wing

another way of raising the temperature of their bodies which gives them a degree of independence of their surroundings. They make use of the metabolic heat generated by their flight muscles which they warm up by shivering their wings before take-off. In this way, the temperature of the flight motor is raised to 30°C or 40°C, which may well be much higher than that of the surrounding air. This action can very frequently be seen in moths when they are disturbed at temperatures too low for flight. The emperor moth, for example, can raise the temperature of its thorax some 20°C above that of its surroundings (P14). The bumblebee, on the other hand, is capable of flight when conditions are only a degree or two above freezing, reaching an internal temperature of over 30°C (P15). For such a tiny creature to generate so much heat requires an enormous expenditure of energy. Even more remarkable is its ability to uncouple the flight muscles from its wings, vibrating and warming them up without actually moving its wings at all.

Since it is essential that insects maintain the temperature of their flight motor, it is hardly surprising that the thorax of most species is given insulation by a dense coating of hairs or scales, as can be seen in butterflies, moths and bees. Even the bare shining thorax of dragonflies is insulated by a layer of air sacs just beneath the outer cuticle. It is also important for insects to take precautions against overheating while flying, as flight muscles that become too hot are unable to function. One of the way insects do this is by passing blood back from the thorax into the abdomen, where it is cooled before being returned once again to the flight muscles.

P14 The large feathery antennae of th[e] male emperor moth (*Saturnia pavonia*[)] capable of detecting the scent of fema[le] from a distance of several miles. Altho[ugh] essentially 'cold-blooded', before taki[ng] it raises the temperature of its flight muscles to some 20°c above that of it[s] surroundings

P15 This bumblebee (*Bombus*) demonstrates the lifting power of insects' wings as it raises its not inconsiderable bulk into the air – note that the wings are twisted through 90° on their upstroke

P16 The caddisfly (*Trichoptera*) couples its hind and forewings together in a similar fashion to moths

Fuel for flight

The energy used by flight muscles during periods of activity is extremely high; in fact, the metabolic rate of the *fibrillar* muscles of flies and bees (see page 46) is about ten times that of the human heart. The fuel and oxygen consumption of this most active of animal tissues is so great that blow-flies may lose as much as 35 per cent of their body weight in one hour's flying.

Fat serves as fuel in the majority of insects, while carbohydrates are used by some others; both have their merits and disadvantages. In terms of weight, fat is far more efficient than carbohydrates, but suffers from the drawback that it has to be broken down and transported from storage to muscles before it can be used. The chief advantage of carbohydrates is that they are soluble in water and are, therefore, able to circulate in the blood that bathes the muscles, and so provide instant energy. But the water solubility of carbohydrates also has certain disadvantages, because insects can only dissolve a limited amount of sugar in their blood, further limiting the supply of potential energy they can carry around. For example, the normal concentration of blood sugar in the honeybee is 2.6 per cent, but once this level falls below 1 per cent, it is quite unable to remain airborne and so is forced to refuel with nectar.

Thus, short-haul insects, like bees and flies, rely on frequent topping up with sugar, while those which have to migrate or fly for sustained periods, such as locusts, burn fat. Although butterflies and moths feed on nectar, they have to convert this into fat before they are able to store it as a source of energy.

The wings

One of the startling dissimilarities between birds and insects lies in the wing structure. All birds' wings are substantially similar, but those of insects exhibit an enormous variety of size and design, ranging from the hair-fringed club-like extensions of minute thrips, to the transparent membranous wings of dragonflies and the scale-covered surfaces of exotic butterflies. Furthermore, unlike birds' wings which are modified forelimbs and so have their own muscular structure, insects' wings contain no muscles or tendons and are merely aerodynamic surfaces controlled and powered from the thorax. Another difference, and one which is frequently overlooked, is that insects' wings are much thinner and flatter than those of birds, and in fact do not resemble ordinary aerofoils at all. Nevertheless, they do act as such because once they begin flapping and the air flows round them, they change shape and become cambered (D38). The nature and extent of this change is not haphazard, but is determined by the varying architecture of the wing which often bends along specific lines or furrows to gain the optimum aerodynamic effect.

D38 An insect's wing becomes a cambered surface when it starts to move

Fully developed wings are only possessed by adult insects, and these grow out of the second and third thoracic segments. They are made up of two layers of chitin which are sandwiched together and strengthened by a network of hollow veins. Most of the wing is dead, but as some of the hairs act as sense organs, the wing is furnished with some nerve fibres. Although many of the veins contain blood, which is particularly important for the wing's development after the insect hatches from its pupa, their chief function is to provide strength. In flight maximum stress occurs around the leading edge, so the veins here are thicker and closer together. The wings of more primitive insects are further reinforced by fine veins crossing the longitudinal ones, but as the tendency of evolution has been to reduce the lateral veining and to strengthen the longitudinal, the veining of the more advanced insects, such as bees and flies, is more simplified. The number and distribution of the veins is faithfully repeated in every individual of a species, and basic patterns are characteristic within groups of related species – an arrangement which is very helpful to the entomologist for the purposes of identification and classification.

Originally, the earliest flying insects had two pairs of wings, each pair functioning independently in flight, a system which has survived up to the present day in some insects such as dragonflies. But generally greater efficiency is achieved when the two wings on each side beat together. Most insect orders have shown an evolutionary trend

P17 A honeybee (*Apis mellifera*) about to alight on the entrance of its hive – the two pairs of 'zippered' wings are half-way through their upstroke

towards a single pair of wings for all practical purposes. Four-winged insects, such as butterflies, moths, bees and wasps, approach a 'two-winged' condition by coupling the forewings and hindwings together. In the butterfly the leading edge of the hindwing has a lobe which locks under the overlapping forewing. Moths link their wings by means of a stiff bristle (*frenulum*), or group of bristles, on the base of the hindwing which hooks onto a catch (*retinaculum*) on the underside of the forewing. The transparent wings of bees and wasps (*Hymenoptera*) are coupled by an elaborate system of hooks on the leading edge of the hindwing, which engages on the forewing's trailing edge (D39).

The beetles (*Coleoptera*) have also followed this trend, but in a different way, as only their membranous hindwings take an active part in flight, the forewings having become modified as protective covers called *elytra*. During flight the elytra are held at about 45° above the body, functioning in a limited capacity as fixed aerofoils to increase the lift (P18). Associated with wing coupling and the two-winged condition, there has been a general tendency for the hindwings to become smaller, which in true flies (*Diptera*) has resulted in

P18 (*opposite*) A ladybird (*Coccinellidae*) taking-off – notice the elytra held at 45°, and functioning to some extent as fixed aerofoils

D39 Wing coupling mechanism of bee

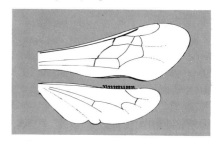

the hindwings being discarded completely; all that remains is a pair of knobbed stalks like drumsticks, known as *halteres*, which help to orientate the insect in flight (P19).

Where insects have found wings more of a hindrance than an aid to their way of life, nature has reversed itself, so to speak, and discarded them altogether. The loss of wings has occurred in species from most orders, including some moths, beetles and flies. The wings of ants would only be an encumbrance in their cramped underground quarters, so while the workers are wingless from the outset, the males and queens rid themselves of their wings immediately after mating. Certain parasitic flies have also lost theirs, as they are hardly needed for rumaging around the fur and feathers of their hosts. The females of some species of moths are wingless, since their lives are spent in laying eggs close to where they themselves were hatched.

P19 A cranefly or daddy-longlegs (*Tipula maxima*) showing the drumstick-like halteres behind the wings

D40 The flight motor of a dragonfly. In the upper diagram a pair of direct flight muscles (coloured) has contracted resulting in an upward movement of the wings. In the lower diagram the other pair of muscles has contracted resulting in the wings' downward motion

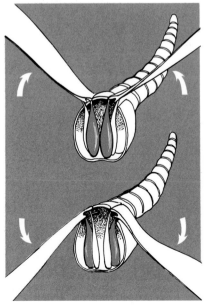

P20 The meadow brown (*Epinephele jurtina*) performing an extraordinary aerial manoeuvre — a multi-flash photograph showing three consecutive images

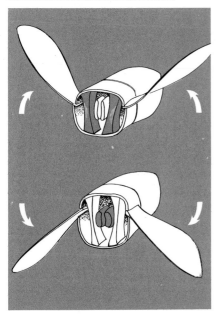

D41 Indirect muscle-powered flight motor, vertical muscles contracted above and horizontal muscles contracted below

The flight motor

The insect's thorax, to which the wings are attached, is packed with a complex of flight muscles, together with various other structures for coupling and operating the wings and, to a lesser extent, the legs. It is from here that all the power and control emanates, enabling the insect to carry out almost any aerial manoeuvre. How the power is applied from the thorax to the wing varies from one order of insects to another, but basically the wings are hinged in such a way that they are free to move in almost any direction, the coupling system functioning like a series of ball and socket joints. The thorax can be considered simply as an elastic box with a lid which is held in place by a membrane coupling it to the top of the sides. The wings pivot on the underside of their bases, and are hinged on to the thorax lid on the upperside. In the more primitive insects, such as dragonflies and grasshoppers, downward movement of each wing is produced by a powerful flight muscle which runs from the thorax floor directly to the base of the wing outside the point of pivoting, so that when the muscle contracts, the wing moves downward (D40). Upward wing movement is powered by another flight muscle which runs from the floor of the thorax to the other side of the wing's pivoting point; contraction of this muscle pulls the root of the wing down into the thorax so that the wing itself moves upwards.

Such an arrangement means that each of the four wings has its own independent power supply provided by the two *direct flight muscles*. In this way, the two wings on

P21 The green lestes damselfly (*Lestes sponsa*) clearly showing the independent movement of the two pairs of wings which are out of phase by about half a cycle

each side need not necessarily beat in unison, as can be seen in the photographs of the praying mantis and damselfly (P21 & 25). The front and rear pairs of wings in damselflies can be, and usually are, out of phase by as much as one half cycle, as is clearly shown in the photograph. The system works very well for insects which do not have a wing-beat frequency much higher than about 25 beats per second, as the nervous impulses which initiate the up and down movements of the wings are also working at the same slow rate.

Many insects, such as bees and flies, have evolved a wing-beat frequency ten times this figure, and as their nervous system is unable to keep pace with such high speeds, they have had to solve the problem differently. To begin with, their thoracic lid is less flexible so that when the flight muscles contract, the whole thoracic box is compelled to change shape as a unit. These insects have two pairs of muscles which, instead of being attached near the wing's base, are connected to the walls of the thorax (D41). One pair, the *indirect vertical muscles*, runs from the roof to the floor of the thorax, while the other pair, the *indirect horizontal muscles*, runs horizontally connecting the rear to the front. The wings are coupled to the sides in such a way that as the thorax changes shape, the wings also move. Thus, contraction of the vertical muscles pulls the lid of the box down and forces the sides out like a squashed rubber ball, resulting in the wings moving upwards; due to the system of leverage, it only takes a small movement at the wing's root to cause the wing to move over a large distance (D41). When the longitudinal muscles contract, the box squashes in the opposite direction, arching the roof of the thorax upwards which causes the wings to move on their downward stroke. It is interesting to note that the muscles that create both the upstrokes and the downstrokes of insects are of the same size, and both deliver power. This is another difference between the flight of insects and birds, as in birds the upstroke is relatively weaker, the muscles which produce it being smaller.

Another modification essential for high-frequency wing movement is the *click mechanism*. The mechanism is a highly sophisticated one requiring an elaborate

P22 Multi-flash photograph of a yellow dungfly (*Scatophaga stercoraria*) as it leaves a rhododendron seed head. The wings of flies are attached to the thorax which acts like a pulsating box: when flies take off, their flight motor is stimulated into activity automatically by special 'starter' muscles connected to the centre pair of legs

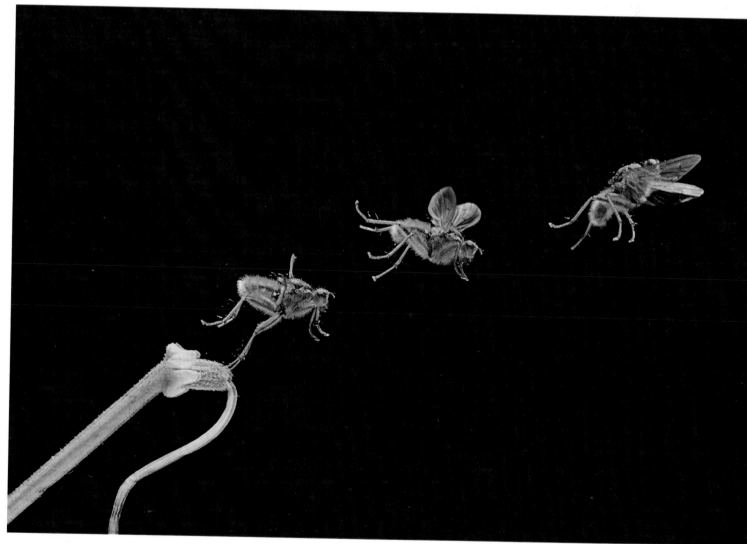

articulation of the wings and special muscle tissue. Its development reaches a peak in true flies (*Diptera*), and bees and wasps (*Hymenoptera*) – the two orders with higher frequencies of wing-beats than any other insects. In simple terms, the wing couplings act as though they were double-jointed with two stable positions, either up or down; thus, they function rather like an electric light switch with an unstable mid-position (D42). As the wings approach their mid-beat position, the sides of the thorax are forced outwards before the wings are able to move fully up or down. Once they have passed this point, the tension is released, and the elasticity of the thorax accelerates the wings up or down to their stable position with a snapping action.

The efficiency of an oscillating system, such as an insect's flight motor, is substantially improved by having high elasticity. This means that it spontaneously resumes its shape after contraction. Once set in motion, the flight motor only needs sufficient energy to replace that which is lost due to wind resistance or is absorbed by friction. The insect's thorax contains many elastic energy-storing tissues, of which the most remarkable is the wing's minute suspension system. This is made of *resilin*, the most nearly perfect elastic material known, and one which so far has defied man's attempts at synthesis.

Together with the development of the click mechanism, insects with high-frequency wing movement have evolved an extraordinary muscle tissue known as *fibrillar* muscle. It is probably the most active tissue ever to have evolved in a living organism, having the peculiar quality automatically of contracting very rapidly after being stretched. Used in conjunction with the click mechanism, once an impulse from the central nervous system initiates the system, the power supply for the wings carries on running by itself until another nervous impulse stops it. Hence, the muscles can contract and relax far more rapidly than the nerves can fire them. In this way, the wing-beat frequency is not really controlled by the insect at all, but depends on various physical factors, such as the surrounding air temperature and size of the insect's wings. It is through a combination of such refined mechanisms that many flies, bees and wasps are able to beat their wings at 200 to 250 cycles per second (1 cycle is a complete up and down movement), while some midges can flap their way through the air at the staggering rate of 1,000 strokes per second.

As well as beating up and down, the wing has to be extensively moved in the other two planes during flight. For instance, the inclination of the wing has to be adjusted so that it can be twisted along its axis. This is accomplished by *accessory muscles* attached to the roots of the wing, and further muscles effect the wing's forward and backward motion. There are also other small muscles which help to control the elasticity of the thorax by bracing the walls against the strain of the main flight muscles.

We often assume that small creatures like insects are simpler than larger ones because they have less inside them; but this is a serious error. Apart from the development of the brain, the fly is almost as complicated and miraculous as man, possessing everything necessary for its special way of life, from an intricate tracheal system to a thorax packed with sophisticated tissues and mechanisms for operating the wings – yet without a second thought, we can destroy it at a single stroke.

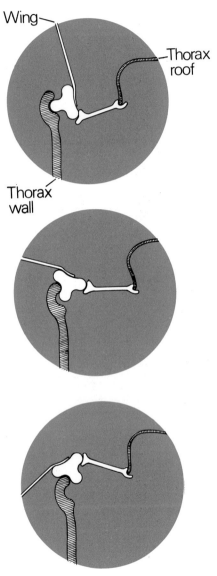

D42 Click mechanism of fly showing wing in mid unstable position (*centre*)

Gliding flight

During normal forward flight the four forces acting on an insect (weight, life, thrust and drag) are in equilibrium (D9). However, when gliding, no thrust is generated; the insect's wings have stopped beating. Now, in order to remain airborne and under control, power has to be derived from the force of gravity (weight), so the airspeed necessary for maintaining lift is obtained by inclining the nose of the insect downwards relative to the air (D43). By so doing, the insect moves forward but steadily loses height in the process; the more efficient the gliding, the shallower the glide path. Although the wings do not flap, minor adjustments have to be made to them to maintain balance and angle of attack.

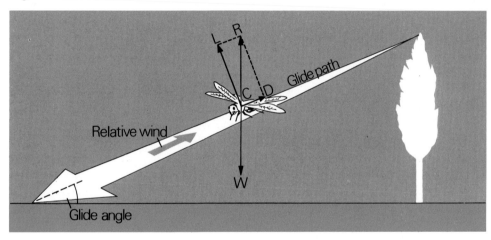

D43 Forces acting on a gliding insect: its weight (W), its lift (L), the drag (D), the resultant upward force (R) together with the glide path and the relative wind

    It is striking that only a few insects have mastered the simple technique of gliding. Some dragonflies, for example, after a period of conventional flapping flight, will sail on motionless outstretched wings for several seconds with little loss in height before resuming normal flight. Some large butterflies, such as the monarch, habitually glide for short distances, and locusts have been know to ascend in thermals to over 3,000 feet with hardly a flap of their wings. These insects glide by relaxing their flight muscles so that the wings lock into an appropriate shallow 'v' attitude (for maximum lateral stability), allowing them to sail through the air without any muscular effort.

Flapping flight

Flapping flight is obviously much more complicated than gliding, and many of its subtleties are still not understood. It used to be thought that insects and birds swam through the sky by pushing their wings against the air, but now we know that this is not so. The wings of both insects and birds act primarily as aerofoils, each section meeting the air at an angle of attack and generating varying degrees of lift, drag and thrust. Such flight depends on a complex combination of wing flapping and wing twisting, and varies according to the species and type of flight.

    Unlike aeroplane wings which only provide lift, those of flying animals have to

P23 A lacewingfly
(*Chrysopa*) caught a
split-second after take-
off winging its way
upwards like a
helicopter

P24 A tree wasp (*Vespa sylvestris*) approaching its nesting hole in a bird box

D44 Path followed by insect's wing-tip

generate thrust as well, functioning more like propellor blades. A propellor drives a blast of air behind it, creating a high-pressure area behind and a low-pressure area in front. It is like a wing to the extent that both lift and drag are generated, but the difference lies in the fact that the lift is directed forwards; nevertheless, the lift is still that part of the force (on the propellor) which is at right angles to the blades' motion.

Whereas the plane of movement of an aeroplane's propellor is at right angles to the machine's horizontal axis, an insect's wing is not confined to any plane. It starts its action from a point above and behind the thorax, and during the downstroke it moves forwards as well as downwards. In reality, it follows an ellipse or slender figure-of-eight, depending on the nature of the flight (D44). The combined effect of these actions is to fan a current of air downwards as well as backwards, providing both lift and thrust, so that the insect not only moves forward but is simultaneously supported against the force of gravity. When the insect wishes to climb, all it has to do is to adjust the angle of its wing-beat so that the current of air is directed more downwards and less backwards.

The flight of an insect is more akin to a helicopter than a fixed-wing aeroplane, in that both propulsion and lift are provided by moving wings. A helicopter in level flight flies with its tail end tilted upwards so that the rotor blades are inclined to the direction of motion in a similar manner to insects' wings in level flight (D45). Both rotor blades and flapping wings move forwards and downwards and then upwards and backwards, both fan a stream of air obliquely downwards analogous to the downwash created by the fixed wings of aeroplanes. During hovering flight the helicopter rotors assume a horizontal plane so that air is drawn from above and flung away downwards. Most insects also use a near horizontal stroke when they hover. For hovering flight see page 94.

The wings of birds and insects achieve a similar effect to the propellor by adjusting their angle of attack with each phase of the wing-beat (D47). Let us follow the cycle of wing movement of a fly. From the fully raised position the wing moves forwards and downwards, accelerating as it does so, attaining a maximum speed in the mid-position. At the start of the movement the angle of attack is extremely high, being about 90°, but this is rapidly reduced as the wing sweeps downwards, reaching its minimum in the centre (D47-C). Once past the central position (quarter cycle), the wings slow down and begin to rotate in the opposite direction with the angle of attack increasing again (D47-D). Thus, the maximum angle of attack occurs when the wing is moving at its minimum speed.

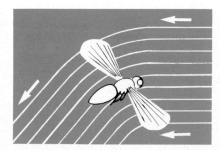

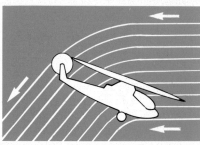

D45 Inclination of insect and helicopter in forward flight and resulting downwash

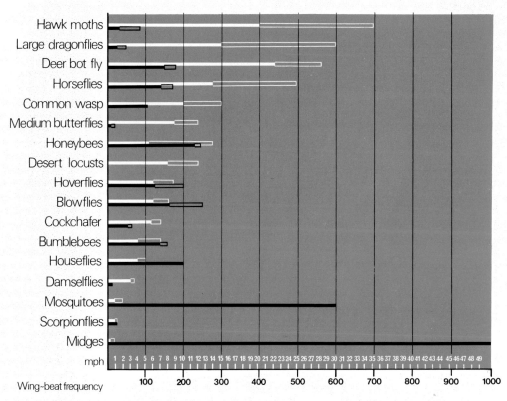

D46 Chart comparing insect airspeeds and wing-beat frequencies

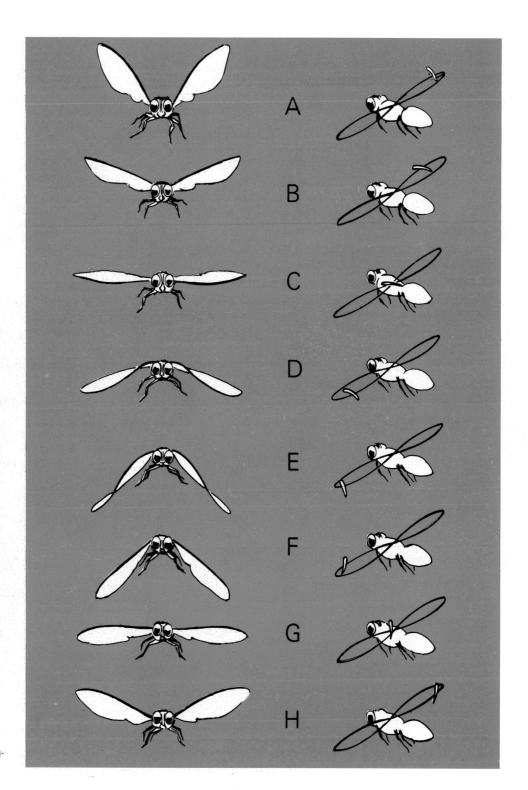

D47 Movement of fly's wings during wing-
beat cycle

At the bottom of the stroke when the speed of the wing is at its lowest but is still rotating, the wing is suddenly moved obliquely backwards and upwards (D47-F). However, it continues to rotate around its longitudinal axis until a stage is reached when the upper surface of the outer two-thirds of the wing may have twisted so much that it faces downwards (D47-G). The degree of this extensive twisting depends on the type of the flight, and is clearly shown in the photograph of a housefly as it takes off (P27). The upstroke continues at an oblique angle so that the wing traces out a path behind that of the downstroke. Towards the top of its movement the direction of rotation changes once again, increasing the angle of attack as the wing reaches the end of its stroke. The complex pattern of movement ensures that the wing attacks the relative wind (remember that the whole insect is travelling through the air) in such a way as to derive suitable proportions of lift and thrust at all stages of its strokes.

The aerial performance of insects varies enormously in terms of wing-beat, speed, endurance and capability, and is largely geared to the requirements of habit and habitat. Although wing-beat frequency can be measured relatively easily with stroboscopes, it is difficult to measure the airspeeds of insects accurately because they rarely fly in straight lines for any length of time. Horseflies are capable of keeping up with vehicles moving at 30 mph over short distances, while some dragonflies are said to be able to do 40 mph over short distances. Although there is no proof, it is quite possible that the fastest insects are the large hawk moths, for they not only have more powerful flight muscles than other insects, but they are also beautifully streamlined. Unfortunately, their nocturnal habits compound the difficulty of measuring their speed accurately. The chart on page 50 shows both wing-beat frequency and approximate airspeeds for various types of insects.

P25 During take-off or when executing erratic turns, the wing twisting of insects can be so extensive that the upper surface faces downwards. The wings of this paper wasp (*Polistes metricus*) have twisted through 180° so that the trailing edges are now leading

P26 The squash bug (*Acanthocephala granulosa*), like many beetles, is a clumsy flier displaying little aerial precision

Flight control and orientation

There can be no doubt that one of the most impressive features of insects is their complete mastery of movement in the air. They seem capable of almost any flight manoeuvre from loops and rolls to flying upsidedown, sideways, backwards, ascending vertically or hovering, or to changing from one to another in a fraction of a second. Watch a dragonfly patrolling a woodland glade as it flies, glides, hovers and changes direction almost instantaneously, sweeping upwards to seize a winged morsel. Observe a swarm of gnats as they hang in the air in their hundreds in an ever-changing pattern of evasive action, never colliding. Or watch a housefly taking off like a rocket and slipping through the air to land upsidedown on the ceiling (P29). All demonstrate sensational flying techniques involving miraculous control.

P27 A housefly (*Musca domestica*) launching itself almost vertically from the top of a loaf of bread

The aerodynamic forces involved for the altering of course are the same for insects as they are for birds (see Chapter 3), but because insects fly at much lower speeds, the methods used are somewhat different. Birds change direction in the air by making adjustments to their tail or wings, the degree of movement needed to make the turn depending on their airspeed. High-speed aircraft require only minute movements of their control surfaces to effect a substantial change in course, but insects, because of their lower airspeeds, have to adopt far more extreme asymmetrical wing positions or movements. Yet the precise nature of these movements is still a little obscure, because insects are so difficult to photograph or film in free flight, especially whilst executing erratic turns. It has been suggested that the wings on each side of the body are flapped at different speeds, but this is highly unlikely, particularly in those insects with indirect flight muscles, as they cannot power the wings independently. Such claims are almost certainly the result of inaccurate interpretation of photographic evidence.

With insects such as dragonflies, the wings can be controlled more or less independently by the direct flight muscles, enabling them to change course by varying the amplitude of the wing-beat on each side of the body. By increasing the amplitude, and hence the power of, say, the lefthand side, the insect turns to the right. But how insects with indirect flight muscles accomplish a differential amplitude is by no means certain; for effecting 'normal' changes in course, small accessory direct flight muscles are used for independent rotational adjustment of the wings at their root. So how are flies capable of those spectacular split-second turns? Recent research in Germany with tethered bluebottles indicates that they may be able actually to disconnect their wings from the flight motor for short periods whilst in flight, allowing one wing to be drawn back into the rest position while the other one beats away normally. The magnitude of one-sided aerodynamic forces produced by these means would certainly help to explain the fly's extraordinary manoeuvrability.

The speed at which flies and other insects are capable of reacting raises a particularly interesting point regarding their time scale. Of all creatures the fly must be one of the most alert, as it is capable of distinguishing accurately between events which are separated by one two-hundredth of a second or less; that vital split-second can be a matter of life or death to a fly as it dodges the hand that tries to swat it. Man is unable to distinguish precisely between events which are separated by anything less than one-twentieth of a second. Our visual system, for instance, allows us to enjoy a film without being aware of the dark intervals between each successive frame, but to a fly the film would appear like a slide-show, with long dark pauses between each transparency.

Aircraft are equipped with a maze of electronic and mechanical instruments for navigation and regulation of their flight. Insects too employ a range of detecting and stabilizing mechanisms for orientation and control, about which little is known. Pitch, yaw, roll and airspeed are monitored by sensory apparatus, such as the eyes, antennae and specialized beds of hairs sensitive to air currents. As far as we know, flies possess the most refined flight orientation equipment in their halteres, which act like oscillating gyroscopes, functioning in a similar way to the turn and slip indicators of aircraft. During flight these oscillate out of phase with, but at the same frequency as, the wings,

P28 In this multi-flash exposure of a shoulder-striped wainscot (*Leucania comma*) taking off, clouds of scales can be seen falling from the insect. A surprising fact came to light as a result of high-speed flash photography: some moths, notably the owlet family, lose wing and body scales in the frantic effort of becoming airborne while their knobbed ends give them an inertia which tends to keep them vibrating in the same plane (P19). When the fly changes attitude or course, the stalks twist slightly as the knobs try to maintain their orientation, and sensory organs at the base detect the degree of flight deviation; in this way, the fly is made aware of its new flight condition. The dragonfly also has developed a similar flight information system by making use of its freely-moving heavy head for detecting differences in the position between it and the rest of its body.

Take-off, landing and stalling

Many insects, particularly those with direct flight muscles, automatically start flying as soon as their feet lose contact with the ground: this is called *tarsal reflex*. Thus, by springing into the air, the wings are stimulated into action. Flies, and possibly other insects, with fibrillar muscles have special 'starter' muscles which run from their middle pair of legs to the roof of the thorax. These automatically stimulate the flight motor into action as the legs kick downwards, giving the thorax a quick tug. Some indication of this action can be seen in the photographs of the dungfly and housefly (P22 & 27).

Before landing most insects extend their six legs out as soon as the landing surface comes within a few body lengths of them. The legs are made to function as efficient shock absorbers, since, unlike aircraft and some birds, insects never run forward after touch-down. They arrive at their landing spot from almost any angle without slowing down at all, alighting with a sharp jolt. Some idea of the efficiency of their landing gear is given by the estimates that some beetles are subjected to a force of about 40 times the force of gravity when they strike an unyielding surface such as the bole of an oak tree – a force which would cause complete disintegration of any aircraft!

Until recently scientists and laymen alike were baffled by the technique used by the fly for landing upsidedown on a ceiling. Does it perform a roll or a half-loop before touch-down? The answer turns out to be both elegant and simple: it does neither, as the multi-flash photograph shows (P29). It flies up towards the ceiling at an angle, stretches out all its legs, using its front ones to touch-down, and then deftly cartwheels over on to its four other feet to complete the landing.

Clearly, insect flight cannot be adequately explained in simple terms, as it is governed by such a vast range of interrelated and coordinated aerodynamic and biological factors. Furthermore, it is virtually impossible to follow, photograph or measure the multiplicity of fluctuating parameters, such as lift, drag and thrust, while insects are flapping through the air in free flight. But over the last fifteen years or so much has been learnt by tethering insects to a highly sensitive balance in a wind tunnel, enabling filming and many aerodynamic measurements to be made with relative ease. Nevertheless, the value of experiments conducted in such unnatural flight conditions obviously has its limitations.

Among the many questions still unanswered is the insect's remarkable immunity to stalling. We know that the critical angle for aircraft is about 15°, but insects do not stall

P29 A fly (*Musca domestica*) does not land on a ceiling by performing a barrel roll or half loop, but by flying up at an angle of about 45° with front feet extended; as soon as contact is made, the fly cartwheels over onto its four other feet

until they reach much higher angles of attack – up to 60° in the fruitfly. Again, unlike aircraft which usually stall dramatically by suddenly falling out of the sky, insects lose lift gradually. This gives them a distinct advantage over aircraft, making them less susceptible to stalling through sudden gusts of wind, and generally allowing them far more freedom in making abrupt manoeuvres. But the way in which insects do this is far from being exactly clear. It is not easy to compare the flight of large bodies with small ones, because flying characteristics change due to differences in scale and surface-to-weight ratio. (As the scale effect is tied up with Reynolds numbers and other complex mathematical relationships, it need not concern us here.)

One critically important factor, and one which certainly affects stalling characteristics, is the behaviour of the thin layer of air in contact with the wing, the boundary layer (see page 23). In insects every vein, hair, scale and corrugation plays a part in the wing's flying efficiency. It has been established that moths deprived of wing scales generate far less lift than those with scales, but other than that we know virtually nothing about the boundary layer's effects on the micro-aerodynamics of insect flight.

P30–34 Various stages of the take-off of a housefly (*Musca domestica*)

P35 Cardinal beetle (*Pyrochroa*) kicking off with its hind feet as it takes to the air. Before a beetle can become airborne it has first to raise its wing-cases (elytra) and then unravel its wings which are folded underneath – this may take several seconds. Note the extensive wing twisting and exceptionally large stroke angle

P36 A praying mantis (*Stagmomantis*) with its legs dangling beneath it shortly after taking off

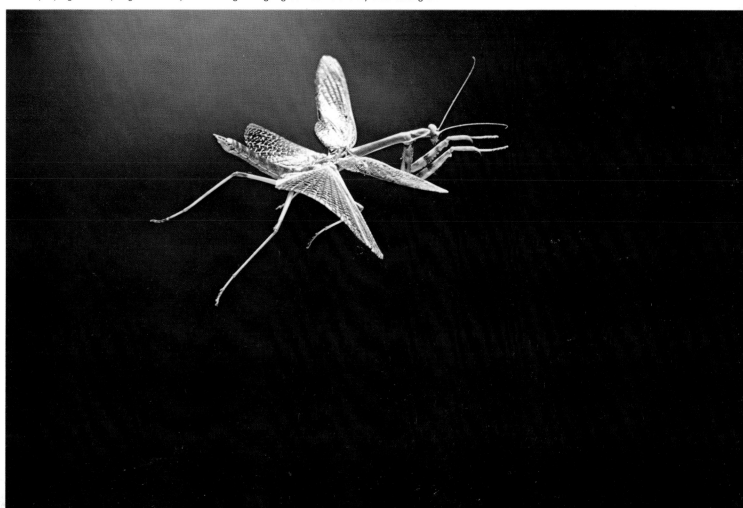

The clap-fling and other mechanisms

The most fascinating aerodynamic discovery of recent times arose from work carried out at Cambridge University by the late Professor Torkel Weis-Fogh. After making a number of aerodynamic calculations of insects, he came to the disconcerting conclusion that certain insects, particularly those with low Reynolds numbers and with low wing-loading, are unable to generate sufficient lift to remain airborne at all, at least according to the accepted principles of aerodynamics. The fact that insects have been flying for some 300 million years must indicate that they derive extra lift by some unknown means. The apparent paradox was taken up by Weis-Fogh who conducted a series of painstaking observations in the laboratory using high-speed cine-photography of hovering insects. Among the insects selected was *Encarsia formosa*, a tiny 1 mm parasitic wasp, often used in the biological control of aphids in greenhouses. As the insect has a wing-beat frequency of 400 cycles per second, the only way to analyze its movements was to film it with a rotating prism camera at 8,000 frames per second.

The first clue to the solution of the mystery was found in a frame-by-frame analysis of the films. It showed that the wasp hovered in the normal manner with its body vertical and the wings sweeping more or less horizontally. However, at the end of the upstroke the two pairs of coupled wings 'clapped' over the insect's back (D48-*A*). Then, after a short pause of one two-thousandth second, the wings were suddenly flung open with their hind margins still touching each other (D48-*B*). Following the 'fling' the hind margins separated and the wings moved horizontally through the air, as in conventional hovering flight.

The *clap-fling* action, as Weis-Fogh described it, occurred in *Encarsia's* every wing-beat cycle, both in hovering and normal flight. Furthermore, the oscillations of the insect's vertical body showed that the lift had equalled the weight of the insect shortly after the clap position, thereby suggesting that air circulation round the wing had been built up a long time before the wings had reached maximum velocity. This is far from what would be expected of ordinary aerofoil action. After further calculations, and

D48 The clap-fling mechanism

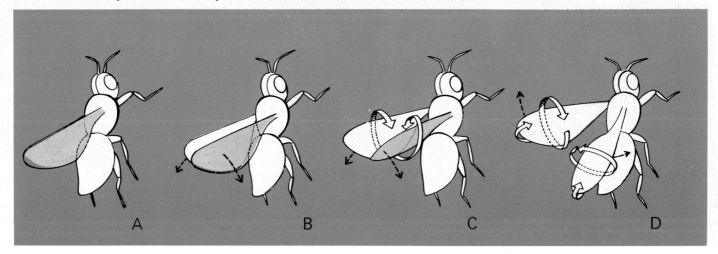

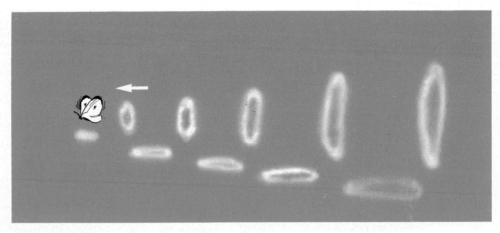

D49 Butterflies leave invisible trails of vortex rings analogous to the wing-tip vortices of aircraft

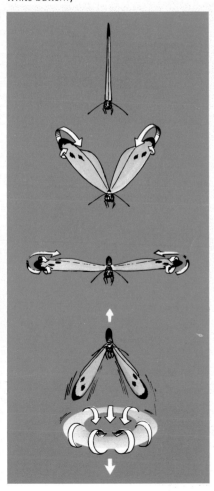

D50 Clap-fling-ring mechanism of cabbage white butterfly

inspired thought, Professor Weis-Fogh found a rational aerodynamic explanation. During the clap the air around the insect is nearly motionless, but as the wings are flung open, it rushes in to fill the growing wedge-shaped space created between the upper surfaces. Thus, as the hind margins separate to begin the downstroke, the wings carry a vortex of air formed during the fling. In this way, circulation and lift are established by the wings as soon as they begin their downward sweep.

Since the discovery of the clap-fling mechanism in this minute wasp, a similar action has been observed in many other insects, including fruitflies, moths, butterflies and lacewingflies (P28, 37 & 38). In the lacewingfly (Chrysopa) the wings are sometimes clapped at the end of the downstroke as well as the upstroke. Even more intriguing is the fact that the lacewingfly's two pairs of wings, which do not necessarily beat together, being controlled by direct flight muscles, sometimes clap 90° (quarter cycle) out of phase with one another.

In the case of the cabbage white butterfly (Pieris brassicae), the clap-fling works in an entirely different way, as is shown by the latest research conducted by Charles Ellington at Cambridge University. It appears that this butterfly has a unique form of flight: it seems to blow a series of 'smoke rings' which are left trailing in its wake (D49). In take-off, hovering and slow flight the downstroke always begins with the wings clapped, but rather than being flung open about the hind margins, they are flung open about the body, which remains nearly horizontal as the wings move vertically downwards (D50). The bound vortices created by the fling, instead of circulating across the wing from leading to trailing edge, move over the wing-tips in a similar way to tip vortices. So, unlike the usual clap-fling or normal aerofoil action, lift is not derived through a Magnus effect, but is generated in the following intriguing way.

At the end of the downstroke the vortices are shed to form one large vortex ring. The fling is used to generate a vortex of air directly beneath the insect, while the force needed to create the ring sustains the insect's weight. Thus, the butterfly does not fly in a conventional aerodynamic way at all, but obtains an upward reaction from the vortex rings which are flung downwards. The effect is analogous to the downwash of normal aerofoil action, but instead of being created in a steady flow, is produced in an

P37 A lacewingfly (*Chrysopa*) taking off;
in the lower image the wings are in the
'clapped' position

P38 A high-speed flash exposure of
1/25,000th second has caught the wings
of this feathered thorn moth (*Colotois
pennaria*) beginning their fling action. As
the wedge-shaped space grows between
the wings' upper surfaces, air rushes in so
that when the hind margins separate, each
pair of wings carries a vortex of air
providing 'early' circulation

P39 Unlike most other hovering animals,
hoverflies beat their wings up and down
through a small stroke angle and keep their
bodies more or less horizontal as in normal
forward flight. As yet no satisfactory
explanation has been found to explain how
they gain sufficient lift while hovering

interrupted sequence. It is interesting to note that the shape of butterflies' wings, with their low aspect ratios and consequent high level of induced drag, is unsuitable for the generation of lift by normal aerofoil action, but is perfect for this function. Moreover, the erratic up and down motion so characteristic of the flight path of many butterflies and moths, particularly those with low aspect ratios and low wing-loading, can now be explained, as the timing of these abrupt movements corresponds to the shedding of the vortex rings.

The discovery of the clap-fling and the 'clap-fling-ring' represents a breakthrough in the understanding of insect flight. There can be little doubt that both mechanisms are far more widespread among insects than has hitherto been observed. No explanation, however, has yet been found as to how hoverflies and dragonflies hover. Unlike normal hovering, these insects beat their wings obliquely up and down through a small stroke angle, while their body axis remains horizontal, as in forward flight (P39). There is no question of any fling action, yet calculations reveal that normal steady-flow aerofoil action is quite incapable of providing them with enough lift to keep them airborne. There must be some other means by which these insects establish air circulation round their wings, but the details are still far from clear.

Although insects have been in the air for millions of years and have witnessed the changing scenes of life on this planet before birds or man learnt to fly, it is only now that we are beginning to appreciate the amazing complexity of their flight. They have taught us that our understanding of the science of aerodynamics was incomplete and have led us to the study of the aerodynamics of non-steady airflow. We are still only at the threshold of knowledge about insect flight, and it would seem that we still have much to learn from nature.

3
The feathered wing

MAN HAS ALWAYS BEEN more intrigued by the flight of birds than by that of any other creature. The relatively rapid and erratic movements of insects quite baffle the human eye and most bats can only be seen at night, but birds' supreme mastery of the air is plain for all to see. The perfect harmony of their form and function has captivated men through the ages, although the subtle blend of innumerable elements which go to make up bird flight has long eluded scientific analysis. Now that the flight of the dove has at last been interpreted and the vibrations of hummingbirds' wings have been recorded on film, the mystery has to a large extent been explained. However, the wonder of bird flight remains.

Birds are the largest class of vertebrates (that is, animals characterized by having a backbone) living today, having diversified into some 8,500 species, compared with the 4,000 species of mammals. The only other living vertebrate to acquire the power of true flight is the bat, although flying fish, a few rodents, marsupials and a lizard are capable of gliding in a comparatively crude manner. Second only to insects, the highest class of invertebrates, birds are the most biologically successful group of animals that have ever existed, and, like the insects, owe their success almost entirely to their ability to fly. Wings and feathers have given birds the power to rise into the air, move rapidly from one place to another, travel long distances, find food not available to other animals, escape from enemies and rear and tend their young in high safe places.

Of all the classes of animals, birds are the easiest to recognize simply because of their feathers. This characteristic has meant that birds, like mammals, are homothermic, or uniformly warm-blooded, which has given them a major advantage over most other creatures. Warm-bloodedness allows an animal to maintain a relatively constant body temperature, usually higher than that of the surrounding air; the blood temperature of birds averages about 5°C higher than those of mammals, including man. In contrast, cold-blooded creatures, such as reptiles, depend entirely on the environment for their body temperature and, although lively enough when warm, the cold makes them too lethargic to hunt for food or escape from enemies.

Warm-bloodedness in both birds and mammals is accompanied by a four-chambered heart which allows fresh oxygenated blood to circulate round the body

uncontaminated by 'tired' blood, as in fishes, amphibians and reptiles. Birds can, therefore, maintain the high rate of metabolism so necessary for intense activity, particularly flight, and because they are able to withstand different climates, can explore all parts of the globe from the polar regions to the equator.

P40 A house martin (*Delichon urbica*) 'braking' and folding its wings before landing on its nest

Evolution

The evolutionary change from cold to warm-bloodedness is at the heart of the question of how birds first came into existence. It seems difficult to believe that birds, whose movements are freer than all other animals, could have descended from the traditionally ponderous and cold-blooded dinosaurs, and yet there is now widespread support for such an ancestry.

The earliest fossil which bears any resemblance to a bird is *Archaeopteryx* (D51). It was unearthed in lithographic limestone by workmen in Bavaria in 1861, two years after Darwin had shocked the world with the publication of his *Origin of the Species*. Up to this time few people had taken his theory of evolution seriously, but the discovery of this most famous of missing links proved almost beyond doubt that birds had in fact descended from reptiles, or at least reptile-like animals. *Archaeopteryx* lived about 150 million years ago in the Upper Jurassic period, during the age of the 'terrible lizards' or dinosaurs, and bore little resemblance to the birds of today. It had many reptilian features, such as a long bony jointed tail and claws on its wings, and instead of a beak it had lizard-like jaws with sharp teeth. As its breastbone had no keel to which powerful

P41 The tree sparrow (*Passer montanus*) displaying its flight feathers as it alights on a garden wall. The feathers of today's birds are structurally indistinguishable from those of *Archaeopteryx* of some 150 million years ago

D51 *Archaeopteryx*

D52 Pterodactyl

wing muscles could be attached, it is generally assumed that the creature was only capable of gliding or at most a feeble flapping. But, in other respects, *Archaeopteryx* was like a bird; it was about the size of a crow, had bird-like feet and was covered not with scales but with feathers which were structurally indistinguishable from those of present day birds.

Because it had feathers, *Archaeopteryx* was undoubtedly a warm-blooded animal and is regarded as the first bird, but over the last few years its ancestry has been the subject of much speculation. The orthodox view has been that it was the thecodonts, a group of meat-eating reptiles of the Triassic period (225 to 195 million years ago) which gave rise not only to *Archaeopteryx*, but also to the pterodactyls and the dinosaurs. But there is an awkward gap of some 60 million years between the thecodonts and *Archaeopteryx*, with practically no fossil evidence bridging the two.

Recently a number of leading palaeontologists have rekindled the theory, much discussed in the later years of the nineteenth century, that birds did not evolve directly from thecodonts, but from dinosaurs. Indeed, there are many who are now convinced that *Archaeopteryx* was not so much a primitive bird as a feathered dinosaur. There are marked similarities between the skeletons of *Archaeopteryx* and certain small dinosaurs, including in one or two examples a fused collar-bone, the wishbone, which is a distinctive characteristic of birds. Even more fascinating was the theory advanced by Professor John Ostrom of Yale University in 1969 that dinosaurs were not cold-blooded reptiles at all, but warm-blooded creatures. Ostrom further believes that instead of the lumbering pea-brained reptiles as they are usually regarded, they may well have been intelligent agile beasts. Even more recently two scientists, Robert Bakker and Peter Galton, have gone so far as to substitute dinosaurs for birds in the five broad categories of vertebrates, hitherto classified as mammals, birds, reptiles, amphibians and fishes. So, in other words, dinosaurs are not extinct after all, but live on as birds.

It is beyond the scope of this book to enter into this debate, except to say that, in the author's view, the theory seems very plausible. The relevant point is that, as far as the evolution of birds is concerned, the crucial development of warm-bloodedness may have already taken place in dinosaurs millions of years before they became birds.

At this point it is worth saying something about pterodactyls (D52). These extra-ordinary-looking creatures which evolved from the thecodonts were, like the dinosaur, probably warm-blooded. They had a continuous expanse of membrane connecting their hind and fore legs in a similar fashion to the wings of bats. Some had a wing-span extending to over twenty-five feet, and clearly tremendous muscles would have been essential for flapping flight, but as there is no evidence that they could have had such muscles, it seems likely that the large pterodactyls did little more than glide. They arrived some time before birds, flourishing for over 100 million years, and then quite suddenly in geological terms, together with the massive dinosaurs which dominated the land, they became the victims of some unknown catastrophe and mysteriously vanished some 65 million years ago. With their disappearance the birds were left to adapt to the changing face of the earth and evolve alongside the insect and the bat.

D53 Early tree-living dinosaur

Once small dinosaurs evolved feathers for insulation, it is possible to imagine how flight may have developed (D53). Running along the ground on hind legs and flapping feathered forelimbs, these prehistoric creatures may have found themselves airborne over short distances, or if, as is more likely, they were tree-living, they may have glided from branch to branch or from tree to ground. In any case, the first wings must have been very crude and no more than aids to gliding, but by flapping and leaping and the gradual process of natural selection whereby their skeletons and general form were modified, they acquired the ability to fly. The early evolution of flight in birds is comparable with that of insects in the sense that they both began with gliding and graduated to flapping flight, as new mechanisms were perfected. Although the early fossil records of birds are incomplete, it is certain that once flight was achieved there was a species explosion among birds, as they rapidly explored new environments and tried new modes of life.

Unfortunately, in the 70 million years immediately succeeding *Archaeopteryx* only a handful of birds are known to have left any fossil remains. Most of these were seabirds, whose chances of fossilization were much better than those of landbirds; the two best known being *Hesperornis*, a six-foot long flightless diving bird with teeth, and *Ichthyornis*, a tern-like bird which had a well developed keel on its breastbone and was therefore most probably a good flier.

Fossilized birds become increasingly more common in the Eocene epoch (54 to 38 million years ago); the remains of flamingoes, rails and game birds have been found in the Paris basin, and the bones of vultures, herons and kingfishers have been dug up in the London clay. One group of fossilized birds found in great number was the giant flightless species, such as the extinct elephant birds and moas from the Eocene, Oligocene and Miocene epochs (54 to 7 million years ago) (D54). These, together with such extant birds as the ostrich, kiwi, rhea and emu, lacked a keel on their breastbone, and so were all considered to represent a stage of development before flight was learnt. At one time all these birds were believed to be closely related, and so were placed in a separate sub-class known as the *ratitae* to distinguish them from the keeled birds. Today it is understood that they are not related to one another at all, but are the result of convergence: that is, where evolution produces similar characteristics within various unrelated species. Each order of these *ratites*, as they are now called, has developed independently in the part of the world where it is found, and probably stopped flying when there was no further need to escape from enemies or obtain food. With disuse their wings and flight muscles gradually atrophied and their keels disappeared.

It is interesting that the loss of the ability to fly occurred in birds on oceanic islands free from predators. Once a bird has lost the ability, it is unable to survive changes in its environment, such as the sudden appearance of predators. Apart from the penguin, which spends most of its time in the sea, there are few flightless birds common today and, furthermore, extinctions over the last few hundred years are relatively much higher in flightless species than those that are able to take to the air. Regrettably, man has been largely responsible for their disappearance, either directly or indirectly; we have only to consider the sad fate of the dodo and great auk.

D54 The extinct elephant bird

Form and function

Flight imposes strict limitations on the shape, size and structure of both flying animals and flying machines. If an animal is to fly, aerodynamic efficiency and power have to be combined with structural and muscular strength, and weight must be kept to a minimum. In birds such requirements have resulted in a certain uniformity of design; for example, the robin has the same basic structure as the pigeon or duck. On the other hand, animals which live on the ground evolve in many contrasting shapes and sizes (compare man, for instance, with a giraffe or a shrew). Every part of a bird's anatomy is perfectly attuned for its life in the air. The compact body of the bird combines strength with extreme lightness unequalled by terrestrial creatures, while its smooth sweeping lines offer minimum resistance to the air.

This raises an interesting, and frequently overlooked, point relating to the startling difference in outward appearance between birds and insects. Insects are generally blunt complicated creatures with appendages and rough structures sticking out in all directions. Birds, in comparison, are streamlined, with pointed front ends and smooth

P42 The tawny owl (*Strix aluco*) coming in to land with its primary feathers and alula fully extended. So that they can approach their prey silently, owls have specially constructed feathers with a fine pile to muffle the noise of the air as it rushes between them

unobstructed contours all over their bodies; even their legs almost disappear amidst the plumage when in full flight, in the same way as the retracting undercarriage of aircraft. Whereas the majority of insects rarely fly more than 10 or 15 mph, birds move at much higher speeds, sometimes exceeding 100 mph. Bearing in mind that the drag at 100 mph is 100 times greater than the drag at 10 mph, then it is clear that efficient streamlining is essential if birds are to attain high airspeeds with minimum expenditure of energy.

The problem faced by birds during their 150 million years of evolution has been more than a matter of aerodynamics and wing development; their whole anatomy and physiology have had to be modified not only to combine maximum strength with minimum weight, but also to improve the efficiency of body functions, such as breathing and blood circulation. Even the senses of birds have been very finely attuned; their eyesight provides them with the maximum possible amount of information at the fastest possible speed, reaching a degree of perfection not found in any other animal. Faced with such specialized changes and refinements it is no wonder that man has failed in all his attempts at muscle-powered flight.

P43 A coal tit (*Parus ater*) streaking from its nest hole at the base of an oak tree, demonstrating the astonishing wing movements of small birds, which may become particularly extensive during strenuous flight activities such as take-off

The effects of these changes can be seen all over the bird's body. To reduce weight the skeleton is remarkably light, the teeth are replaced by a light horny beak and even the reproductive organs almost disappear once the breeding season is over. To accelerate the rate of fuel combustion the blood temperature has had to be raised. The metabolism is also helped by an unusually large heart which beats at a fantastic rate: about 600 beats a minute in the robin and 1,000 beats a minute in the hummingbird. Birds also have a unique way of breathing, for as well as the usual pair of lungs found in all other warm-blooded animals, they have air sacs spread around their bodies within muscles, abdominal organs and even in the hollow cavities of their bones. The sacs are connected to the lungs and windpipe by a complicated arrangement of tubes and shunts. On inhalation most of the air bypasses the lungs and flows into a series of posterior air sacs, while the remainder, after passing through the rear of the lungs, flows forward into another set of anterior air sacs. On exhalation the air flows out of the posterior air sacs through the lungs and out of the windpipe, while at the same time the air in the anterior sacs flows directly out of the windpipe (D56).

Although the essential exchange of oxygen and carbon dioxide takes place in the lungs, the lungs are supplied with oxygen in such a way as to provide a continuous flow into the bloodstream. The system is known as *cross-current* flow and is outstandingly efficient, not only for normal flight, which is strenuous enough, but particularly for high-altitude flying where the air is rarified. Many birds regularly fly at over 5,000 feet on their migrations and some have been plotted on the radar screen at over 20,000 feet. Cross-current flow explains how they survive flight at such altitudes.

Unlike insects which are held together by an exoskeleton, birds, in common with other vertebrates, have bones. Bones not only give the body form and shape, but also provide rigid points to which muscles are attached. In the majority of animals the skeleton is the heaviest part of the body, but many of the bones of birds are honeycombed with air spaces. Additional strength is imparted by a criss-cross of internal bracing struts, in a similar way as the frame of an aeroplane. This *pneumatization*, as it is called, is particularly apparent in the upper arm bone, or *humerus*, of gliding and soaring birds (D55).

Several bones which were once separate in birds' remote ancestors have fused,

D55 Longitudinal section of upper arm bone of a bird showing pneumatization

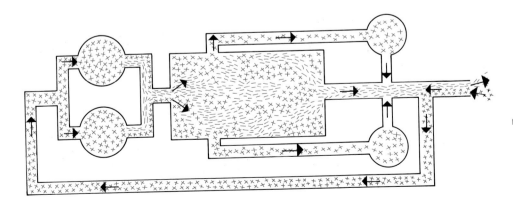

D56 Schematic diagram illustrating principle of cross-current flow of bird's respiratory system. Air enters the windpipe from the right into the lungs (the central rectangular area) via the posterior air sacs (left). The anterior air sacs are on the right

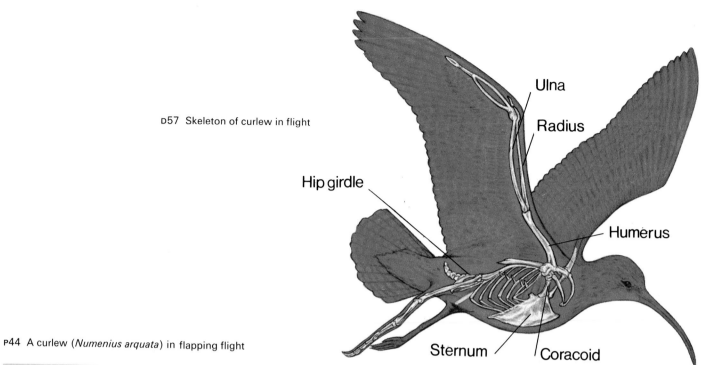

D57 Skeleton of curlew in flight

Ulna

Radius

Hip girdle

Humerus

Sternum

Coracoid

P44 A curlew (*Numenius arquata*) in flapping flight

reducing to some degree the weight of bone present, and also strengthening such regions as the skull, chest and pelvis (D57). For instance, the hip girdle is constructed of a fused plate of bone so that it can withstand the shock of landing and the stresses imposed during the spring in take-off; it also provides a strong surface for the anchorage of leg muscles. The two collar bones, or *clavicles*, are also fused at their bases forming the wishbone, and together with another bone, the *coracoid*, help to brace the wings and prevent the body from being squashed when the powerful flight muscles contract. Since extensive neck movement is essential for feeding, preening and as an aid to all-round vision, the neck bones of birds are markedly different from those of other animals. Whereas nearly all mammals only have seven neck vertebrae, birds have as many as twenty-five, giving them greater flexibility.

Perhaps the most highly modified bone of all is the breastbone, or *sternum*, which has to anchor the bird's massive flight muscles. For this reason, it has a deep keel jutting out of its centre; generally, the deeper the keel, the more powerful the bird's flight. Naturally, the wing bones too have been rigorously adapted to meet the requirements of flight; the largest bone is the humerus, which is short and thick and bears the main flight muscles which are connected at their other end to the keel of the breastbone. The humerus is connected to the forearm which is made up of two bones, the *ulna* and the *radius*. The inside bone of the forearm, the ulna, provides a foundation for the secondary flight feathers, while the wing-tip, which is analogous to our hand, comprises a series of finger bones and carries most of the primary feathers largely responsible for propelling the bird through the air. The bone that corresponds to the thumb provides a fixing point for the *alula*, a small tuft of feathers for reducing the stalling speed.

P45 Contour feathers

P46 Having fed its young this kingfisher (*Alcedo atthis*) backs out of its nest hole in the bank of a stream and deftly executes a 180° 'reverse flip' into the air

Feathers are exclusive to the bird and distinguish it from all other creatures; they make up the entire wing surface supporting the bird in the air. The aerodynamic form, lightness and remarkable ability to change shape of the feathered wing represents perfection in functional design. In spite of the saying 'as light as a feather', the plumage of many birds sometimes weighs twice as much as the entire skeleton. The origin of feathers is still a mystery, yet they have played a vital role in the bird's evolution. By the time *Archaeopteryx* appeared they were fully developed, and have remained virtually unchanged ever since. It is assumed that they started as a primitive and fluffy form of insulation, and gradually evolved into the highly complex structures possessed by *Archaeopteryx*. Feathers provide insulation in the same way as the hair on mammals, by holding layers of air between the animal's skin and its surroundings, so helping to maintain the body at an even temperature. They also mould the body contours from bill to tail into a smooth streamlined shape and are important for waterproofing, display and camouflage.

Feathers are of two main types: the down feathers, or *plumulae*, and the varied outer flight and contour feathers, or *pennae*. The down feathers lie underneath the cover of the contour feathers and their chief function is to provide insulation (P45). The beauty and complexity of their structure can only be fully appreciated when viewed under the microscope. The vaned feather is made up of a central shaft which is hollow up to about two-thirds of its length. Beyond this point, known as the *rachis*, it becomes solid to increase the strength at the thinner end. Growing out from the central shaft are hundreds of *barbs* which make up the *web* of the feather. Each barb in turn carries hundreds of tiny filaments called *barbules*, and these are equipped with minute hooks which interlock with the adjacent row of barbules (D58). The whole structure works much like a zip fastener. If the web splits as a result of the hooks becoming disengaged, the bird simply draws the feather through its beak a few times and the condition of the web is restored; this is one of the purposes of preening.

Preening is an essential spare-time activity of birds and, together with periodic moulting, provides an effective all-the-year-round system of feather maintenance so vital for the bird's health and survival. The flight and tail feathers need special attention because, if they become worn or damaged, their aerodynamic efficiency is impaired. But all the rest of the plumage has to be kept clean, dry and in the right place, and as some large birds may possess as many as 25,000 feathers, restoration can become an endless chore. The chances are that if a bird is not feeding, sleeping, courting or nesting, it is perched up in a safe place preening itself. Feathers have a clear advantage over the chitinous wings of insects or the wing membranes of bats and the extinct pterodactyls: instead of becoming seriously and irretrievably torn, damage is more like to be confined to a smaller area and can be repaired. Furthermore, if for some reason the damage is severe, feathers can be replaced by the process of moulting and regeneration.

Power for flight is supplied by muscles which function by burning fuel with the help of oxygen carried around with the blood. In the first instance, food is broken down into suitable carbohydrates and fats by the bird's efficient and rapid digestive system; the difference in colour between various types of muscles provides a clue as to the nature of the fuel used. Game birds, such as chickens and pheasants, have white flight muscles (as in the 'white meat' of the breast) which are fuelled by *glycogen*, a carbohydrate which provides the instant energy requirements for rapid emergency take-offs and short bursts of strenuous flight necessary to these birds. Because white muscle fibres operate anaerobically, that is without oxygen, and the waste products accumulate rapidly, they soon become fatigued and need resting after use. This explains why pheasants and grouse are unable to summon enough energy for taking off once they have been flushed out a few times in quick succession. On the other hand, the flight muscles of pigeons and other long-distance fliers burn fat which requires a rich and constant supply of blood to provide sufficient oxygen for extended periods of activity. Consequently, these flight muscles contain a large number of capillaries, and so are much darker in appearance. Some muscles, such as those in the breasts of eagles and other soaring birds, are an in-between colour since they contain both types of muscle tissue, generating their energy by burning both fat and glycogen.

Only about a fifth of the fuel is used for working the muscles, the remainder is lost in

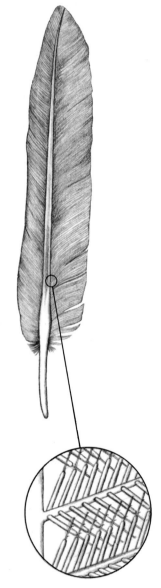

D58 A flight feather with a magnified section showing the structure of the web

the form of heat, some of which has to be removed to prevent damage to the body. Birds circulate blood through their muscles, thereby transferring the heat elsewhere in the same way as most engines lose their excess heat by circulating water through a radiator. Whereas bats are able to lose excess heat by passing blood through the tiny membranes of their wings, birds carry it to the lungs where it is lost by the evaporation of water at the lung surface. If the system becomes overloaded, panting can help to cool it further, or if this is still not sufficient, the bird can evaporate water from the floor of the mouth in what is called *gular flutter*.

When a bird takes to the air, a large number of muscles come into play, of which the most powerful are the two pairs of breast muscles responsible for the up and down motion of the wings. Between them these muscles constitute a large proportion of the weight of the body, which in strong fliers, such as pigeons, may account for more than one-third; consequently, they are positioned around the bird's centre of gravity to help stability of flight. By far the largest flight muscle is the *pectoral* which produces the downbeat, the wing's power stroke; it is anchored in between the keel of the sternum so that when the muscle contracts, the wing moves down. A smaller breast muscle, the *supra-coracoideus*, pulls the wing up again; it is attached to the sternum in a similar way to the pectoral at the lower end, but the tendon at its top is connected to the upper side of the humerus and is looped over the top so that, when the supra-coracoideus contracts, the wing moves upwards (D59).

Wings, of course, do not simply flap up and down – they also have to be operated and twisted into all sorts of positions about the joints in the shoulder, elbow and wrist (D60). To effect and control these movements there are numerous other muscles within the wing; even feather follicles have their own muscles so that they can be raised, lowered or moved sideways to assist in flight manoeuvres. The complexity of birds' wings accounts

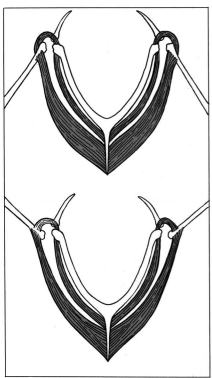

D59 Action of bird's breast muscles showing pectoral muscles contracted above and supra-coracoidii contracted below

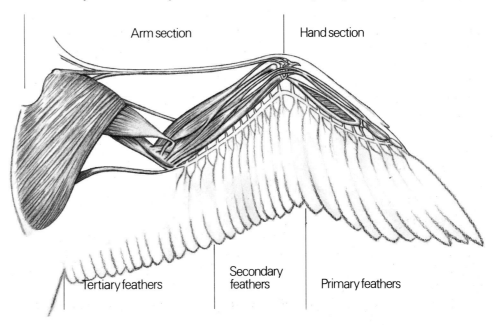

Arm section   Hand section

Tertiary feathers   Secondary feathers   Primary feathers

D60 Muscles and flight feathers of a bird's wing

P48 & 49 Wren (*Troglodytes troglodytes*) carrying insects to its young, seen poking their heads out of the domed nest. Notice the slotted primaries of one wing and the propellor action of the other wing

P47 As the kingfisher (*Alcedo atthis*) approaches its nest carrying a fish, it slows down by inclining its body to the airflow. To reduce stalling speed the alula on each wing has been extended

for one of the major differences between the flight of insects and birds. As we have seen in the previous chapter, the power control of insect wings is effected from within the thorax. Whereas insect wings are for all intents and purposes flat plates which only assume certain aerofoil characteristics once they are in motion, birds' wings are examples of perfect aerofoils. The bones in the leading edge make their wings rigid and rounded, and the feathers at the trailing edge taper to a point; the efficiency of the aerofoil is further improved by a hollowing of the wing beneath.

The *alula*, sometimes referred to as the bastard wing, is an interesting refinement which acts as a subsidiary aerofoil in front of the leading edge of the main wing. Under normal flight conditions the alula is folded back out of the way; but when the airflow over the upper surface becomes turbulent as the bird approaches stalling speed, the alula is spread forwards forming a slot through which air rushes, restoring a smooth fast airstream and curtailing stalling (D61). The device functions in the same way as the Handley-Page slot built into some aircraft (see page oo). The alula's action, particularly at low airspeed when the bird is approaching its nest, can be clearly seen in the photographs of the starling and the kingfisher (P52, 53 & 47). Many slow-flying birds, such as buzzards and crows, make use of the same principle by having deeply slotted primary feathers, so that the outer section of their wings becomes a series of narrow aerofoils allowing the air to slip through the spaces, reducing turbulance even more (P51). Whereas soaring birds have almost permanently slotted wing-tips, all birds are capable of spreading their outer primaries to a greater or lesser extent for reducing stalling when landing, as shown in the photographs of the tawny owl and the wren (P42, 48 & 49). It is worth mentioning that insects lack these obvious slow-flying devices, as they can fly slowly enough without them.

Gliding flight

Compared with a fixed-wing aircraft or even insect wings, a bird's wing is an incredibly complicated structure of muscles, tendons, blood vessels and nerve tissue, which, by a combination of muscular movements and feather bending, is capable of flapping and changing its shape in a bewildering variety of ways. The subtleties of wing movement are still not fully understood, but bird flight in its simplest form, that is gliding, does not require flapping or the consumption of muscle power. Simply by stretching out the wings, the outer part merges with the arm section to form a continuous plane, so that the bird flies in a similar manner to a fixed-wing aircraft.

Many species have so exploited and perfected gliding that they are able to maintain height and travel long distances without flapping their wings at all. Some birds soaring at high altitudes indulge in all kinds of aerial displays: ravens, wood ibises and eagles have frequently been seen flaunting their powers of aerobatics, including loops, barrel rolls and inverted flying, in a way which can only be interpreted as exhibitions of pure exuberance and joy (D62).

All birds are capable of gliding to a limited extent, even those birds with high wing-loading, such as pheasants or grouse, once they have propelled themselves into the air by a burst of vigorous flapping. Many birds are able to glide for longer periods if the wind is strong enough to provide the necessary lift. In normal winds birds, like insects, glide using the force of gravity to maintain a suitably fast airflow over the wings by inclining the glide path downwards. In perfectly still air relatively little progress would be made as the bird would soon reach the ground, but in fact the atmosphere seldom remains static for rising bodies of air are frequently encountered in the form of thermals and wind currents.

Thermals are caused by the uneven heating of land-masses, when rising columns of warm air expand into huge bubbles. As the bubble ascends, currents within it produce a central ring of revolving air, rather like a smoke-ring, with a flow of colder air rising through its centre (D63). If the thermal is strong enough, a soaring bird, although flying

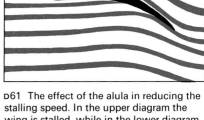

D61 The effect of the alula in reducing the stalling speed. In the upper diagram the wing is stalled, while in the lower diagram the slot formed by the extended alula has restored a smooth airflow

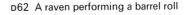

D62 A raven performing a barrel roll

downwards relative to the air, is able to ascend while circling on the updraught within the bubble. The bird must be able to circle within the diameter of the thermal's vortex ring, where the upward airflow is strong enough to sustain its weight. As the bird turns, some of its lift is applied to another force, *centripetal force*, in keeping the bird to a circular path and preventing it from sideslipping. The tighter the turn, the greater the loss in lift, which means that the rate of sink increases as the radius of the turn becomes smaller. At the same time, the lower the wing-loading, the lower is the sinking speed and the smaller the radius of turn. So, naturally, those birds with low wing-loading and small size can exploit smaller and weaker thermals than the heavier and larger species. If the thermal becomes too weak, the bird flaps away to find some other thermal which provides better support. In this way, the bird can ascend in one thermal, then glide or fly to an adjacent one and make long cross-country journeys with minimum expenditure of energy. Gliding of this kind is called *static soaring*.

P50 Black-headed gulls (*Larus ridibundus*) gliding into the wind at the same speed as it blows them backwards

P51 The turkey vulture (*Cathartes aura*) is a master of thermal soaring. Note the deep slotting at the wing-tips for minimizing turbulence at slow airspeeds. It was the turkey vulture which gave the Wright Brothers the clue to lateral control in aircraft

Landbirds, such as eagles, buzzards and vultures, are masters of the thermal and are often carried to many thousands of feet. Vultures rely almost entirely on these rising bubbles of air for flight and, for this reason, are most commonly found in tropical countries where thermals are plentiful. In the early morning before the sun has warmed the ground, vultures make little attempt to fly. The smaller species begin to leave their perches as soon as the weaker thermals start forming, but the larger vultures have to wait before taking to their wings until the sun gets high enough for the stronger

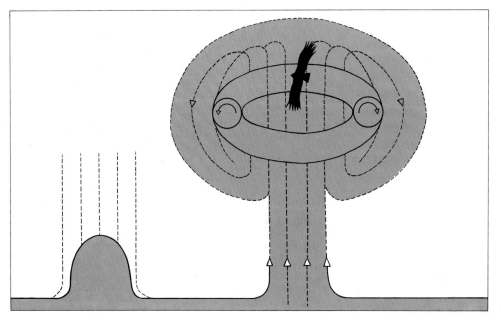

D63 Formation of thermal

P52 A starling (*Sturnus vulgaris*) swinging
its body into a near vertical position and
spreading its tail to reduce landing speed.
Its claws are poised ready to clasp the bole
of the willow tree where it has its nest

P53 To reduce stalling speed, the starling's
alula is spread forwards from the wing's
leading edge

thermals to develop. Birds which soar at high altitudes have special characteristics; so that the rate of sink can easily be offset by the rising column of air, soaring birds have large deeply cambered wings of low-loading to provide them with maximum lift (P51). Their slow-flying performance is improved by a well-developed alula and deeply slotted wing primaries which smooth out turbulence and reduce stalling speed to a minimum; extra stability and manoeuvrability are provided by a large tail.

As a way of life it may seem the height of ecstasy to soar effortlessly on motionless wings in the hot sun, gazing down on the world below, but to vultures it is a matter of survival. Carrion feeders, such as vultures, need to soar to heights of several thousand feet with minimum effort to survey with the help of their extraordinarily acute eyesight a vast area of countryside for suitable food. Soaring enables them to travel long distances by gliding from one thermal to the next so that they can carry food back to thier chicks from up to a hundred miles away.

There are many other birds capable of remaining airborne without using muscle power; seabirds, in particular, exploit wind currents to maintain or increase height, and, like the broad-winged soarers, frequently travel long distances by means of this economical form of locomotion. Seabirds make use of updraughts, but of a different nature. These may be caused by various factors as, for instance, when an on-shore wind hits a cliff and is deflected upwards (D64). Even when the wind is off-shore, currents of air will curl upwards as the wind spills over the cliff's edge. Gulls and other seabirds take advantage of these updraughts; fulmars are especially fascinating to watch as they wheel to and fro, sometimes hanging motionless, near the cliff-tops.

It may seem surprising that the open sea can provide suitable conditions for gliding, or *dynamic soaring* as it is called. When winds blow over water, the lower layers of air are slowed down by friction with the waves. As a result, a gradient of velocities is produced with the wind speed at its maximum at 100 feet or so above the surface. Albatrosses exploit this situation to the utmost (D65). After an upwind climb to about 100 feet, they turn to face downwind and enter a dive, gaining airspeed on the way and covering

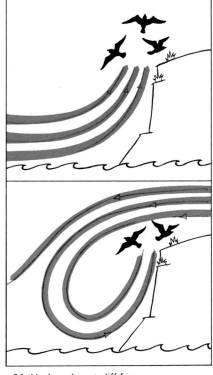

D64 Updraughts at cliff face

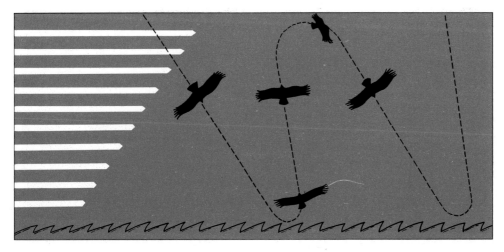

D65 Dynamic soaring at sea

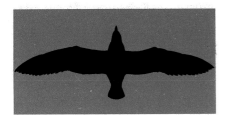

D66 A silhouette of an albatross showing its high aspect ratio

considerable distance. At sea level they turn into the wind using their momentum to climb again into stronger winds, during which time they progressively lose speed. But this again is made up for by the gain in height necessary for the long downwind glide. Once they can climb no further, they turn about and the cycle is repeated. The albatross spends most of its life out at sea, many species living in the southern regions of the Pacific Ocean where the wind sweeps almost continually from west to east. Here, these large birds fly hundreds of miles with scarcely a flap of their wings; in fact, recent recoveries of ringed specimens prove that they are capable of travelling round the world – some 19,000 miles at 40° latitude – in eighty days by riding the wind. When on the rare occasions the wind speed drops to zero, albatrosses abandon flight altogether and settle on the water to await the next breeze.

Oceanic birds owe their gliding skill to the special shape of their wings and bodies. Like the static soarers, they have a large wing area, but instead of the heavy build of the vulture family, gulls and albatrosses have slender streamlined bodies and long narrow wings to keep form and induced drag to a low level. Such a shape is far more efficient for sustained high-speed gliding, when loss of height has to be minimized (D66).

Flapping flight

In spite of the gliding ability of a number of well-known birds, the vast majority are quite unable to soar or glide for more than a few yards, and even soaring birds are unable to do so all the time. This means that power for flight has to be supplied by the bird itself flapping its wings. Flapping flight in birds, broadly speaking, is similar to that of insects in that the wings of both have a dual action, working not only as aerofoils, but also propellors, so that lift and propulsion are effected by a complex combination of vertical and horizontal movements and spiral twisting. In birds, as in insects, these movements vary according to the species and for different types of flight: for instance, the wing movements of a swan in level flight are not the same as those of a blue tit, while the pattern of wing movement made by a pigeon in fast forward motion is entirely different to that of its take-off flight.

Each section of the bird's wing can be visualized as an aerofoil simultaneously oscillating in the three planes of space about a pivot which is in itself moving forward, while at the same time the overall shape of the wing is perpetually changing – a much more intricate arrangement than the wing of an aeroplane. During its complete cycle of movement the wing-tip moves further and faster than any other part of the wing, generally following a complicated figure-of-eight pattern in a similar fashion to insects. Due to their high airspeed the primary feathers provide most of the lift and propulsion, and as these feathers are so flexible, the degree and direction of their bending is a valuable guide to the direction and magnitude of the aerodynamic forces involved: for example, an upward bending indicates the force of lift and a forward flexing one of propulsion. Evidence of these forces at work can be seen from most sharp photographs of birds in active flight.

Functionally, a bird's wing can be considered in two sections: the hand section which

P54 Barn owl (*Tyto alba*) leaving its nest hole after feeding its young. The articulation between the hand and forearm section of the wing is clearly shown

bears the primaries, and the inner part from the forearm to the shoulder which carries the secondary and tertiary flight feathers. The articulation of the two sections is clearly shown in the photograph of the barn owl (P54). In fast forward flight the movements of the thicker inner part are mainly in the vertical plane and provide a significant proportion of the lift. When the pectoral muscle contracts, the wing which is fully extended moves downwards, forcing the air down which induces an upward reaction on the bird (D67-A–C). At the same time, further lift is generated by the wing's natural aerofoil properties as it moves through the air.

The movements of the outer part of the wing are more extensive and complex, with the wing-tip following an involved and tortuous path. During the downstroke the outer section moves slightly forwards as well as downwards, and so that the wing encounters maximum air resistance, the primary feathers overlap one another forming an airtight surface (D67-C). Thrust is provided at the wing-tip by the outer primary feathers which twist as they are forced through the air. Each flight feather is so designed that the web is wider and more pliable at its rear edge than at the leading edge. Consequently, air pressure resulting from the wing's movement bends the trailing edge more easily so that each feather resembles the blade of a small propellor. As the primary feathers bend back during the downstroke, the wing's leading edge twists increasingly downward from the base to the tip so that the wing becomes propellor-shaped, with the most pronounced twisting towards its weaker end, the tip (P68 & 71). In this way, the air is driven backwards and the bird is propelled forwards.

In level flight most of the upstroke is one of passive recovery, requiring little or no effort by the supra-coracoidii muscles. As the wing moves up, it is rotated about the shoulder so as to increase the angle of attack. At the same time, the outer wing is partly folded and the primary feathers twist open rather like a venetian blind, allowing air to pass between the spaces so that minimum resistance is offered (D68). The effect is clearly shown in the multiple-flash photograph of the little owl (P55). In small birds with relatively fast wing-beats the upstroke is largely neutral in its effects. Heavy birds, however, with slow wing-beats and higher wing-loading are less able to afford a 'wasted' stroke, particularly when flying slowly or taking off. To increase the power of the upstroke they have relatively large supra-coracoidii muscles; and to increase the forward drive the end of their upstroke is usually punctuated by a backward flick of the wing-tip. In so doing, they derive extra thrust as a result of the air operating against the back surface of the primaries (D67-E). Certainly, pigeons adopt this technique in forward flight as is shown by high-speed photography, and it is reasonable to assume that many other birds must do the same, but to what extent is still uncertain.

Small birds with high wing-beat frequencies have more extensive wing movements than large species. This is dramatically shown in the photographs of a coal tit as it streaks from its nesting hole (P43). Slow flight in all birds, as is usually used for take-off and landing, also demands a much higher degree of wing movement and twisting, reaching a maximum when the bird hovers in still air, thus enabling much higher levels of lift and/or propulsion to be generated than would otherwise be possible. The differences between the two types of flight can be seen by comparing D67 with D69. In slow flight the

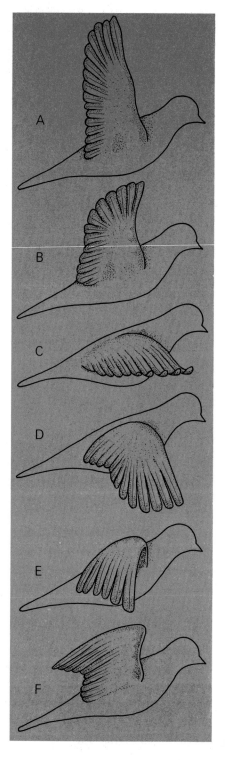

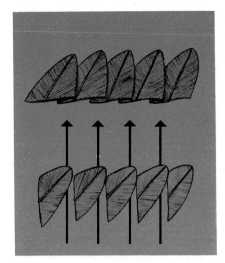

D68 Venetian blind effect of primary feathers

D67 (*left*) Action of pigeon's wing in fast forward flight

D69 Action of pigeon's wings in slow flight

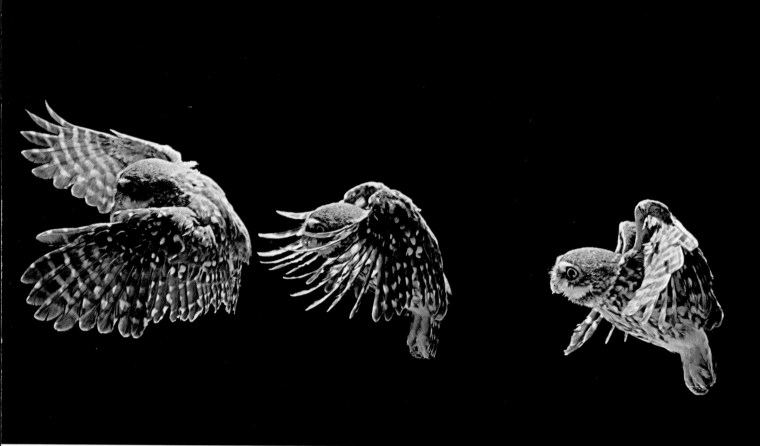

P55 A multi-flash photograph of a little owl (*Athene noctua*) coming in to land. As the owl approaches its selected landing spot from normal horizontal flight (right), the bird slows down by increasing the angle of attack of its wings and inclining its body to the airstream. At the same time it spreads its tail which acts as an airbrake, and gradually lowers its undercarriage. Also notice the venetian blind effect of the wings' primaries for reducing the air resistance during the upstroke

P56 A swallow (*Hirundo rustica*) darting through a two-inch space between the stable door and the beam above. Notice the asymmetrical tail and wing position as the bird executes an abrupt turn

bird's body is inclined to the horizontal, the angle of attack is greater, the alula and primaries are opened and the twist at the wing-tip during the downstroke is increased – all of which can be seen in many of the photographs in this book. While in both slow and fast flight the wings are fully extended at the start of the downstroke, in slow flight the amplitude of movement is far greater, the wings starting their descent at about $90°$ above the horizontal and practically touching at the end of the downstroke (D69-D). Furthermore, the wings move much further forward relative to the shoulder, frequently stretching a long way in front of the bird's head. Due to its low airspeed the inner section of the wing contributes little or no life or propulsion in normal slow flight; it is the tip section that generates most of the aerodynamic forces.

As soon as the upstroke begins, the upper arm rotates around its axis so that the wing is thrown into a vertical position with the tip section flexed at the joint and primaries elevated above the bird (D69-E & F). At this stage, the whole wing is drawn violently backwards and upwards and is simultaneously extended (D69-G & H). The movement can be compared with the use of the backward flick of the wing-tip already described. But in slow flight the movements involved are far more exaggerated enabling, among other things, birds to accelerate into the air, so speeding up their transition into normal forward flight. It is highly unlikely that there is any sharp division between fast and slow flight, the pattern of wing movement gradually merging from one to the other.

As well as generating lift by standard aerodynamics, there is now strong evidence to suggest that pigeons, like insects, make use of non-steady airflow to provide them with extra lift. As early as 19 BC Virgil observed in the *Aeneid* that rock doves made loud claps with their wings when taking off in a steep climb. In 1890 a Frenchman, E. J. Marey, in his study of animal locomotion, *Animal Mechanism*, demonstrated that the sounds were caused by the bird actually clapping its wings above its back. Since then high-speed photography has proved that when a pigeon takes off in an emergency, immediately preceding the downstroke the wings are fully extended and almost touching above its body. This is followed by a powerful pull of the pectoral muscles, resulting in the wings' rapid downward acceleration. The whole movement is indicative of a true clap-fling mechanism (see page 62), particularly as the lifting force generated by this sudden movement is clearly shown by the degree of bending in the primary feathers well before the wings have reached their horizontal position. Furthermore, calculations made by Professor Weis-Fogh indicate that under these conditions the wings of pigeons begin their movement with an air circulation and resultant lift of a magnitude that otherwise could only be produced much later in their down-stroke.

It, therefore, seems almost certain that pigeons, and probably many other birds, make use of the clap-fling mechanism in a similar way to insects for generating extra lift when flying at low airspeeds, such as those encountered when taking off or hovering. Some evidence that the wings of the kingfisher may possibly be half-way through a fling is indicated in plate 3. Notice that the wings' trailing edges are in contact with each other, and have clearly been subjected to considerable aerodynamic force as the bird dives headlong out of its nest-hole in the bank of a stream.

Hovering flight and steering

Hovering flight, like slow flight, is strenuous, expensive in terms of energy and requires pronounced wing movements (P57 & 72). Although many birds can hover for a second or two, only a few species are capable of sustained hovering; hummingbirds use it most often while feeding, and the kestrel and tern use it as a substitute for a perch while hunting for food. True hovering flight can be defined as flying while remaining in one spot in still air, and under these conditions the wings generate lift without propulsion. The action of the wings in hovering varies according to the species of bird and its type of flight. One form of hovering, which relies on the wind and requires little effort is practised by some larger birds, such as buzzards, which fly into the wind at the same speed as it blows them backwards, so that in relation to the ground they remain stationary.

The hummingbird is the supreme virtuoso of hovering flight among birds, although in hovering it resembles bees and hawk moths more closely than birds. Its wing movements are of considerable amplitude, swinging downwards and forwards in the downstroke, and upwards and backwards in the upstroke. At the same time, the bird's body is inclined at a steep angle to the horizontal, which allows the wings to move horizontally through the air like the rotor blades of a helicopter (D70). Indeed, the propellor action of the relatively long primary feathers of the hummingbird is essentially the same as the action of helicopter blades, in that airflow is directed downwards producing lift, rather than backwards as in normal forward flight. Apart from hovering, hummingbirds can fly upwards, downwards, sideways and even

D70 Hummingbird in hovering flight showing horizontal path followed by wing-tip

backwards, their breathtaking ability to manoeuvre allowing them to dart up to a flower, hover in front of it for a moment to sip the nectar, back away and dash nimbly to the next.

It is of utmost importance to birds to be able to manoeuvre easily in the air; this they can achieve without inherent stability, the two being mutually incompatible (see Chapter 1). Whereas full-scale aircraft are designed to provide some form of compromise between the two, birds do not need to be inherently stable as they can stabilize themselves by reflex action. When a bird's flight path is disturbed by turbulence in the air, the movement is detected by the semi-circular conals in its inner ear and compensations are made by involuntary adjustments to the wings and tail.

A bird steers in the air by deflecting the airflow to one side or the other, adopting asymmetrical wing positions, and tilting the angle of its body or using its tail as a rudder (P56). For straight and level flight the position and movements of each of its two wings must be identical so that the forces acting on them are also identical. In making a turn, the forces are assymmetrical: for instance, a turn to the left is executed by

P57 An arctic tern (*Sterna macrura*) in hovering flight (also see P72)

P58 The greater spotted woodpecker
(*Dendrocopos major*) diving out of its nest
hole. The wings are half-way on their
downward beat – notice the extended alula
of the near wing

deflecting the airflow to the right, while at the same time banking to the left, deriving more lift from the righthand wing. In this way, the bird is subjected to an external force (centripetal force) acting towards the centre of its turning circle, but to avoid inward or outward sideslipping, the force has to be related to the bird's weight, airspeed and radius of turn. As the centripetal force for the turn is derived from part of the lift, it means that to compensate for this loss, the angle of attack of the wings during the turn has to be increased to obtain sufficient lift. Thus, the effective wing-loading during a turn is always greater than in level flight, and for this reason there is always a higher risk of stalling when making tight turns.

The rate at which a bird is able to change direction in the air depends on the relationship between its weight and the airspeed and the area of control surfaces brought into play to produce the turn. At high speeds only a slight touch to the wings or tail is necessary to effect an abrupt change of direction. Therefore, fast fliers, such as swifts, can manoeuvre easily and do not need large control surfaces, consequently they have small tails. On the other hand, slow-flying birds, such as buzzards and other static soarers, have large tails to give them flexibility and stability, and a pheasant uses its long tail for twisting between the branches as it rises from its woodland habitat. Swallows, which are generally fast on the wing, but also capable of flying slowly in a surprisingly controlled manner, have forked tails which are closed for high-speed flight, but are spread out when flying slowly (P56).

Take-off and landing

Take-off and landing are the two most testing moments in flying. Both aircraft and birds have to make a smooth transition from one medium to another. On take-off the basic problem is to obtain enough lift for leaving the ground in as little space as possible, while on landing it is more a question of maintaining control at low airspeeds.

When an aircraft roars down the runway, it seems incredible that the machine will ever leave the ground; but with birds, particularly the smaller species, the movement appears effortless and almost instantaneous. Indeed, the apparent ease and rapidity with which these creatures rise into the air and attain high speeds is one of the most remarkable features of bird, as well as insect, flight. Some of the disparity in the take-off between aeroplanes and birds can be explained in terms of scale, birds being so much smaller and lighter, but the main difference can be attributed to the higher aerodynamic efficiency of flapping wings at low airspeeds. We know that when birds are in normal fast flight, a substantial part of the lift is obtained from aerofoil action, but at take-off there is no forward speed, so the initial airflow has to be provided by other means, such as running, jumping or the extensive flapping described earlier.

Most birds take off from the ground or a perch by springing into the air with their powerful feet to gain initial impetus followed by the strenuous slow-flight wing flapping, the powerful downstroke providing most of the lift, and the upstroke most of the thrust (P60 & 68). As the smaller flight muscles, the supra-coracoidii, supply the power for the upstroke, they would soon become exhausted, so flight of this strenuous

D71 Silhouette of pheasant

D72 Silhouette of swift – note widely contrasting aspect ratios

nature cannot be sustained for long periods; once the bird has built up speed, the pattern of wing movement reverts to normal. The amount of effort sometimes required for take-off is suggested by the way a pheasant explodes into the air (D71). Its deeply cambered short broad wings enable it to extract immense lift in a flurry of intense activity, which is maintained only for a few seconds.

Some large birds do not have a special take-off flight at all; instead they use a similar technique to aeroplanes. Swans, albatrosses and divers (P59), for example, simply turn into the wind and run along over the surface of water or the ground, flapping their wings until they gain enough airspeed for lift-off. The hummingbird, instead of pushing off with its feet, creates so much vertical lift with its wings that the bird pulls the perch up before letting go. Obviously the easiest way for a bird to become airborne is to acquire the assistance of gravity by dropping into space from a height. Seabirds, such as puffins and guillemots, plummet down from cliff ledges, and swallows and martins can launch themselves into the air from the sides of buildings or their nests. Some species, such as swifts, are so built for high-speed flight that this is the only way that they can become airborne – once they become grounded they are only able to take off again with the greatest of difficulty (D72).

Whereas take-off normally requires power and the expenditure of energy, the transition from air to ground involves considerable skill and judgement. Young birds frequently overshoot their perches and crash into the undergrowth in their early

P59 The red-throated diver (*Gavia stellata*) taking off from a Scottish loch

P60 A Louisiana heron kicking-off from an Everglades swamp

P61 Like the Concorde, swans such as this mute swan (*Cygnus olor*) need to drop their noses to improve forward visibility when landing at high angles of attack

attempts to land. Birds, like aircraft, have to reduce speed for landing so as to lessen the chances of injuring themselves. This can be done in a number of ways: large heavy birds try to do it by landing into the wind, but such a technique is not necessary for the smaller species. Most birds lose speed by swinging their bodies into a near vertical position and spreading their tails which act as an airbrake, so that they meet with maximum air resistance (P52 & 55). At the same time, the angle of attack of the wings is progressively increased so as to maintain lift as the airspeed is reduced, while the alula and slotted wing-tips are brought into play to keep stalling speed as low as possible. More drastic braking can be effected by flapping in reverse, in the same way as a jet aircraft uses reverse thrust, so that the airstream is directed forwards rather than backwards.

Birds, such as pigeons and woodpeckers, lose speed by dipping down some distance before their selected landing spot, completing the final few feet in an upward glide, so as to reduce the momentum of their flight; the final impact of landing is absorbed by the legs in the same way as an aeroplane's undercarriage. Waterbirds, particularly those with webbed feet, benefit from the cushioning effect of the water as they skid to a halt (P61). Although most waterbirds tend to land somewhat rapidly, the albatross has an unusually high stalling speed, which means that a brisk oncoming wind is a great advantage for slow approaches on to its nest. When there is no wind, the bird has no option but to crash-land, the feet absorbing the initial impact and its well padded breast taking the rest of the impact as it flops forward.

Birds have often been described as living aeroplanes and certainly their flight, particularly when gliding, is more akin to a fixed-wing aircraft than the helicopter action of many insects. It is, however, interesting to speculate whether, without bird flight as a model, man's conquest of the air would have been long delayed or perhaps even now be a goal beyond his reach.

4

Evolution of manned flight

SINCE THE DAWN OF HIS HISTORY MAN could not have failed to be impressed by the mysterious phenomenon of flight, as birds and insects were constantly around for him to observe and admire. He must have longed for wings of his own so that he too might take off and experience the rush of cool air as he swooped and turned in the skies. Unobstructed and free, he would be able to escape from the constraints of his earthbound existence. It is no wonder that wings with their power, mobility and speed have always symbolized the unearthly, the superhuman and the divine.

The image of the winged object has been worshipped all over the world for thousands of years, and it appears in the legends, fables and religions of many nations. The mythologies and literature of both the Orient and the West abound with winged supernatural personages as well as legendary creatures with wings in the form of dragons, vampires, griffins and jabberwocks. In Egypt the soul was pictured as a human-headed bird, and in Greece the psyche was a butterfly; but when evil forces were depicted, their leathery wings were modelled on those of the innocent, but nocturnal, bat. Although legends are full of witches on flying broomsticks and magic carpets, there can be little doubt that man's deeply ingrained desire for wings is largely due to his envy of the bird. The bird has not only been his chief source of inspiration to fly, but also a model on which his own efforts to do so could be based.

Fact and folklore

Nobody knows exactly when, where or how the first attempts to convert those dreams into reality were made, for it is here that myth and truth become confused. It is not always possible to differentiate between the actual and the imaginary early flights. Some were unquestionably fanciful enterprises, but perhaps it does not matter so long as the underlying theme on which all these tales are woven is recognized: the tantalizing lure of the sky and man's constant longing for wings.

As the desire for flight stemmed from an emotional rather than a rational urge, it is not surprising that the early attempts lacked any scientific basis. In any case, systematic scientific thought was still unknown; man had no conception of lift or aerofoil action. It seemed that his only line of approach was to try to emulate the bird, which he did, as well as attempting a few other intriguing and bizarre experiments. The most celebrated

P62 Daedalus and Icarus, from a woodcut
of 1493

of these is encapsulated in the legend of Icarus, whose father, Daedalus, made him a
pair of feather wings held together by wax. He flew so close to the sun, however, that the
wax melted and sent him crashing to his death in the sea below, while his less ambitious,
but horrified, father flew safely across the Aegean Sea to Italy. One of the earliest, and
more likely, endeavours to take to the air was reputed to have been in about 1500 BC by
a Persian king, Kai Kawus, who tethered four starving eagles to his light aloe-wood
throne; he encouraged them to fly by suspending legs of lamb on spears above each
corner. The famished eagles are said to have raised the throne and its pilot a short
distance before giving up the hopeless struggle.

In early historical times the development of aeronautics also remains very vague. We
know that aeronautical devices in the form of arrows and boomerangs had been in use
since the earliest days of civilization; strictly speaking the arrow is a simple projectile
with stabilizing feathers, while the boomerang is a power-launched rotating glider. The
aerodynamics of the boomerang are so extraordinarily complex that it baffles the
imagination as to how primitive tribes were able to develop such a device. Kites too are
of great antiquity, dating back to around 1000 BC, and yet it was not fully understood
how they worked until the last century. They originated in China where they were used
not only as toys, but also for warfare, and over the centuries before the development of
the aeroplane huge man-carrying kites were employed for spying and dropping fire-
bombs over enemy territory. In the nineteenth century George Pocock, an English
schoolmaster, constructed a large kite for pulling his horseless carriage around the
countryside at astonishing speeds, much to the bewilderment of onlookers. Although
kites are actually tethered gliders, they were not seriously considered as a potential
means of aerial locomotion until the nineteenth century, and played little part in the
early history of aviation.

Ornithopters and Leonardo da Vinci

There are very few reliable records of man's attempts to fly until early medieval times when the so-called 'tower-jumpers' appeared in force. Among these many courageous, but perhaps misguided, men was a twelfth century Saracen of Constantinople, who provided himself with a voluminous cloak and leapt off a tower; he crashed to the bottom and killed himself. In 1503 Giovanni Danti, an Italian mathematician, after fixing wings to his arms, jumped from a tower at Perugia; he was seriously injured. Four years later John Damian, Abbot of Tungland and physician at the Scottish Court of James IV, attempted to fly with wings from the battlements of Stirling Castle. Needless to say, he too fell to the ground and broke his thigh.

Leonardo da Vinci in the fifteenth century was the first man to give the problem of human flight serious consideration. His genius in this field, as in many others, is unquestionable. Driven by insatiable curiosity, he was obsessed by all flying creatures, and birds, in particular, were the objects of his constant observation. His profound understanding of anatomy enabled him to study the structure of bird's wings, and observe their movements both in flapping and soaring flight. He tried hard to understand the relationship between wing movement and its effect on air currents and to investigate pressure, centre of gravity and streamlining. He soon realized that bird flight could not be thoroughly understood without a knowledge of the air and its forces. Leonardo not only recognized the force of inertia, first scientifically expressed by

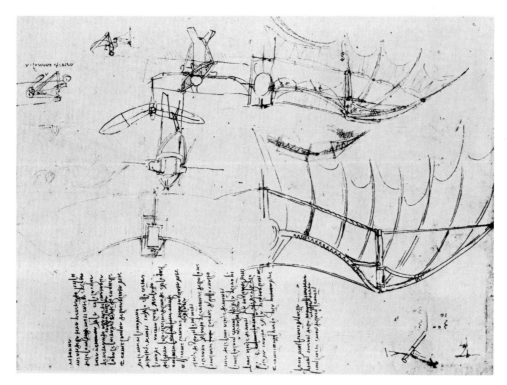

P63 Leonardo's glider taken from his notebooks, about 1485–1500

P64 (*opposite*) Leonardo's studies of bird flight, also taken from his notebooks

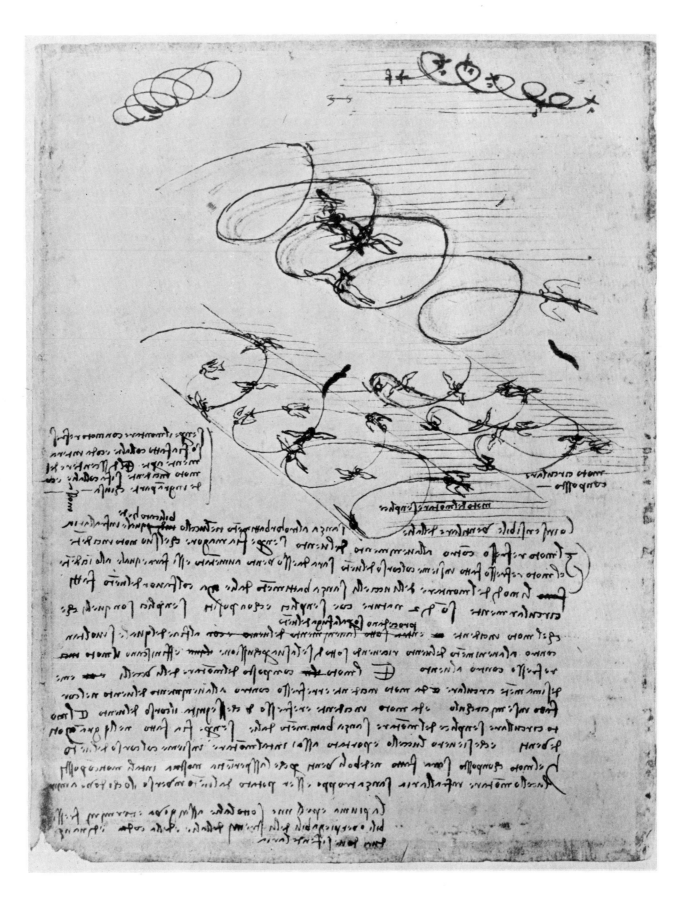

Galileo a hundred years later, but also foreshadowed Newton's third law of motion. He observed that 'the movement of the wing against the air is as great as that of the air against the wing.' He made copious notes and drawings of all these investigations in the notebooks which he kept assiduously for some forty years, and from these it is clear that he was preoccupied with the idea of a mechanical imitation of the flapping flight of birds

P65 Leonardo's designs for ornithopters showed that he intended them to be powered by a combination of leg and arm movements activated by means of intricate machinery

and bats. Consequently, much of his work concentrated on flapping-wing aircraft, or *ornithopters*; but in applying himself to this end his genius proved misguided. One reason for this was that he had compared bird flight with swimming and oarsmen, and had erroneously deduced that birds 'row downwards and backwards'. Such a conclusion seems hardly surprising considering the speed at which birds vibrate their wings; even the slower actions of eagles and gulls were too fast for accurate analysis which was only made possible by photography.

Leonardo's designs for ornithopters show that he intended them to be powered by a combination of leg and arm movements activating extremely intricate and heavy machinery. It is as well that a full-sized model of one of his contraptions was never tested, since a flight based on backward-flapping wing movements would have been disastrous both for pilot and machine. Human muscle power, moreover, would have proved totally inadequate for propelling his cumbersome machines; even today, with extremely light strong materials, it seems unlikely that sustained man-powered flights will ever be a practical possibility. Leonardo proposed an alternative way of powering his ornithopters involving a bow-string mechanism which had to be rewound by the pilot in flight. He also suggested a machine with fixed wings, to whose outer ends were attached panels hinged for flapping. Curiously enough, it was not until after he had spent all his energies on ornithopters that he wrote his extraordinary treatise on bird flight, *Sul Volo degli Uccelli* (1505), but in it he was still unable to discover the secret of bird propulsion.

It is a pity that in all his works Leonardo clung so tenaciously to the concept of the flapping wing, neglecting the fixed-wing glider, which only appeared later in his notebooks in the form of thumb-nail sketches. One of these drawings shows a man clinging to a flat board seen in gliding descent, an idea which, if he had developed it, could have led to a vital change in his views of flight and to more fruitful conclusions.

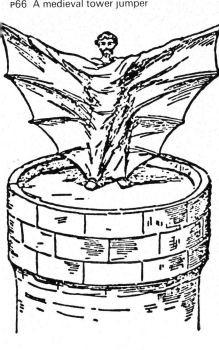

P66 A medieval tower jumper

As well as designing flying machines, Leonardo was the first to consider the principle of the parachute and the helicopter and to design a head-harness to operate an elevator intended to control an aircraft's rate of climb and descent. Yet in spite of his painstaking efforts in recording all his experimentation and research, his work remained more or less unknown for more than three hundred years until the end of the nineteenth century. Thus, tragically Leonardo's aeronautics failed to influence the future history of flying; if his researches had been published sooner, as he had intended, manned flight might have taken a very different course.

After Leonardo's death in 1519 nearly a century passed before any other thinker gave flying further serious attention, although the tower-jumpers continued their frightening escapades well into the seventeenth century. For instance, in 1673 in Frankfurt-am-Main, a brave gentleman by the name of Bernoun broke both his neck and legs when he tried flying with wings, and a French tight-rope dancer by the name of Allard was seriously injured when attempting a flying display before Louis XIV. In 1663 the Marquis of Worcester published an account of his attempts to make a ten-year-old boy fly from one end of a barn to the other; the aeronautical details of this event are far from clear. It was unfortunate that the concept of human muscle-powered flight was unwittingly perpetuated by Francis Willughby (1635–72), one of the first British naturalists to treat the study of birds as a science. By calculating the strength of the pectoral muscles of birds and comparing them with the human arm, he concluded that any hopes of flying would depend on the leg muscles, rather than the arm muscles, being harnessed to power the wings. This encouraged the development of unworkable contraptions based on leg movements.

Two seventeenth-century scientists, however, Robert Hook (1635–1703) in England and Giovanni Borelli (1606–79) in Italy, quite independently came to the conclusion that human muscle-powered flight was impossible, and claimed that unless man could decrease his weight or increase the size of his muscles, flapping flight would be an impossibility. Today we understand that it is simply a question of power-to-weight ratio; man is not designed to fly, he is heavily built with large solid bones and weak breast muscles. By contrast, many of the bones of birds contain air, while their breast muscles are enormously powerful when compared with those of man. Even if we attached highly efficient aerofoils to our arms and legs, we would still be quite incapable of powered flight. It has been calculated that in order to house suitably powerful muscles for flight, humans would require shoulders six feet broad; as Robert Hook observed, an independent source of power was essential.

Despite the deterrent findings of Hook and Borelli and the ridicule and contempt to which most would-be aviators were subject, man's irrepressible ambition to fly continued to manifest itself in the form of ornithopters and flapping wings powered by the human frame. Perhaps the most celebrated of these performances was made in Paris by the Marquis de Bacqueville, who in 1742 fastened wings to his arms and legs and, watched by a large crowd, leapt from a house in an attempt to fly across the Seine; he floundered miserably down onto a washerwoman's barge and broke both legs.

As an alternative to wings, the seventeenth and eighteenth centuries also saw various intriguing and mostly wholly impractical schemes to get man airborne. A Jesuit priest, Father Francesco de Lana de Terzi, published a book in 1670 which included a design for an aerial ship that was to be suspended by four copper spheres emptied of air. He even gave his aerial boat a sail, not knowing that his craft would be unable to obtain advantage from the wind, as a free balloon must inevitably drift with the wind in any case. He also failed to appreciate that the atmospheric pressure would have collapsed his spheres; nevertheless, his idea represented the first conception of a lighter-than-air craft. An even more ingenious mode of aerial locomotion was suggested by Cyrano de Bergerac (1619–55) for lunar travel: observing that dew rose when sunlight fell on it, he proposed to strap round his body bottles of dew which he reasoned would be soaked up by the sun and hence carry him up too.

One of the most absurd contraptions was the *Passarola*, the invention of the Portuguese Bartolomeu Laurenço de Gusmão (1686–1724), and until recently treated as a work of pure fantasy. As it appears in prints it resembled something between a balloon-cum-boat and a bird-powered parachute, but there is reason to suppose that this Heath-Robinson device may have been tested. Although the *Passarola* itself could never have flown, evidence now suggests that de Gusmão did build a model which made

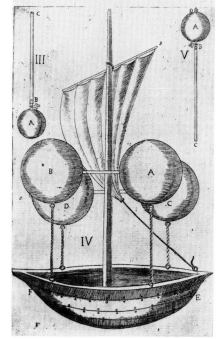

P67 Francesco de Lana's aerial ship of 1670, taken from a contemporary engraving

P68 To gain its initial thrust this starling (*Sturnus vulgaris*) has used its feet for kicking off from the bole of a tree. Even if man had sufficiently large wings, he would still need breast muscles six feet across to propel him through the air

a tentative flight before the king of Portugal in 1709. A contemporary report strongly indicates that it functioned as a primitive hot-air balloon in so far as it got off the ground at all; thus, while Archimedes conceived the principle of flotation, it was de Gusmão who two thousand years later may have been the first to demonstrate lighter-than-air flight.

Balloons, Cayley and modern aeronautics

There can be no doubt that the invention of the balloon during the eighteenth century had a profound effect on the progress of winged flight. Balloons not only provided a rival method to winged flight for aerial navigation, they could be used as testing rigs for parachutes, propellors and aero-engines. It is strange that for thousands of years scientists had failed to realize the implications of forest fires and volcanoes when

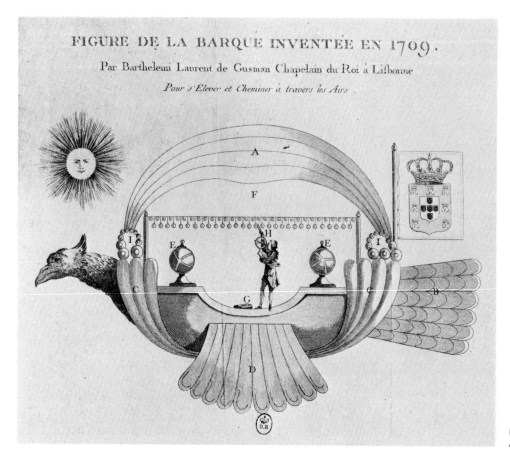

P69 An engraving illustrating Gusmao's *Passarola*, about 1700

fragments were carried to great heights by the rising column of hot gases. This discovery was the work of a French paper-maker, Joseph Montgolfier, who in 1783 while contemplating by the fireside was struck by the lifting powers of hot air. Without understanding what made it rise, he carried out an experiment with a bag made of fine silk under which he lit a fire; much to his delight, and to that of his landlady, it filled out and rose to the ceiling. In June of the following year Joseph and his brother, Étienne, demonstrated for the first time before a large crowd in the market-place of Annonay a large model hot-air balloon. Five months later in Paris man's first aerial journey took place in a *Montgolfière* and shortly afterwards, also in France, the first man-carrying hydrogen balloon was released. Thus, after centuries of striving in which countless men had lost life and limb with flapping wings and other fantastical machines, made in total ignorance and disregard of laws of physics, man floated into the air; at last he could become airborne at will.

Following the discovery of ballooning there was curiously little progress, or even interest, in winged flight until the beginning of the nineteenth century, when one of the greatest figures in the history of flying appeared on the scene. The invention of the balloon in 1783 fired the imagination of Sir George Cayley, who was only ten years old

P70 Man's first aerial journey in a
Montgolfier hot-air balloon in 1784, as
shown in a contemporary engraving

P71 The wings of the stock dove (*Columba oenas*) are half-way through their downstroke. Note the propellor action of the primaries as the wing's leading edge twists increasingly downwards from base to tip

at the time. From that moment he devoted his whole passion and energy to the theory and practice of flying until his death in 1857. He was the first man of science to apply his mind to the fundamental principles of mechanical flight, publishing his seminal work, 'On Aerial Navigation', in 1809. This paper, in which he wrote, 'Aerial navigation will form the most prominent feature in the progress of civilization', laid the foundations from which grew all subsequent developments in aviation. This and his later works marked the turning point in man's long struggle to acquire his own wings; historians now consider that Cayley was the true inventor of the aeroplane.

By reviewing everything that was already known about air pressure and ballistics and by conducting his own investigations into air resistance, Cayley was able to formulate the new science of aerodynamics. Like Leonardo, he looked to the bird in the hope that it might provide some clue, and his observations show him to be the first man to come to grips with the basic technique of bird flight. He soon discovered that birds, once airborne, expend far less energy on flying in a straight line than on taking off. Further laboratory experiments on the resistance of air to inclined surfaces moving through it led him to the conclusion that a bird in full flight somehow derived its support from the air itself, while its flapping wings were used mainly for propulsion. (He was even able to calculate that a gliding rook would be capable of maintaining height when its airspeed reached 25 mph.)

He also recognized the potential lifting power of a wing in discovering that the upper surface had a region of lower pressure than that on the underside; from this fact he surmised that a cambered aerofoil might generate more lift than a flat surface. In his paper, 'On Aerial Navigation', he summed up the whole problem of mechanical flight thus: 'To make a surface support a given weight by the application of power to the resistance of the air.' So, at last lift and propulsion had been separated, and the way was open to the solution of winged flight by man by the application of power to fixed wings. Borelli had been proved right and the imitation of flapping birds' wings should now have been discarded for ever.

P72 The arctic tern (*Sterna macrura*), like the kestrel, uses hovering as a substitute for a perch when hunting for food. Hovering is the most strenuous form of flight and requires extensive wing movements

Most of Cayley's practical work was done at Brompton Hall near Scarborough, Yorkshire, and it was here in 1804 that he made the first successful aeroplane in history. It was a five-foot-long model glider based on the kite; the main plane was fixed, while the tail unit was attached by a movable joint and acted as an elevator and rudder control. A movable weight was also provided for adjusting the centre of gravity. Cayley had already recognized the importance of stability and his aeroplane gave the first practical demonstration of how it could be achieved. In the summer of 1809 when he was thirty-six years old, he built a full-sized glider which he tested unmanned and which also flew a short distance carrying a boy. Cayley recorded that it would frequently lift him up, and convey him several yards.

After designing a variety of ingenious flying machines of more historical interest than practical value, Cayley returned later in his life to gliders. In 1853 he constructed a full-sized triplane with built-in longitudinal and lateral stability; it also had control in the form of an elevator-cum-rudder operated by the pilot. It was in this machine that his reluctant coachman was persuaded to fly across the valley at Brompton and so make the first gliding flight in history. Full details of the design of this machine were published in a popular contemporary magazine; unbelievably, no one in the scientific world took any notice of it at all. Like Leonardo da Vinci, Cayley was certainly a genius, but much more practical in his approach to flight. He anticipated in almost every way the aeroplane of a century later and expressed his firm conviction that 'this noble art will soon be brought home to man's convenience, and that we will be able to transport ourselves and families and their goods and chattels more securely by air than by water and with a velocity of from 20 to 100 miles per hour.' Sadly, in spite of his accurate predictions and the publication of his brilliant research, his work was largely ignored, even by the Aeronautical Society which was set up in London in 1866, until some twenty years after his death.

Serious interest in mechanical flight was rare in the middle years of the nineteenth century. One reason for this was that the secret of flotation (Archimedes' principle) was far simpler to comprehend and apply than the aerodynamic principles governing winged flight. Various other factors combined to postpone exploration along the lines suggested by Cayley, such as the unsuitability of the steam engine and the continuing preoccupation with ornithopters and human muscle power.

Even while Cayley was alive, the subject of winged flight was regarded with scepticism and ridicule by the public, which was hardly surprising since there were all manner of eccentrics, showmen and charlatans about who did little to add dignity to the science of aviation. The 'flappers' still seemed as determined as ever to master the skies

P73 Cayley's model glider of 1804; from a photograph of a modern replica

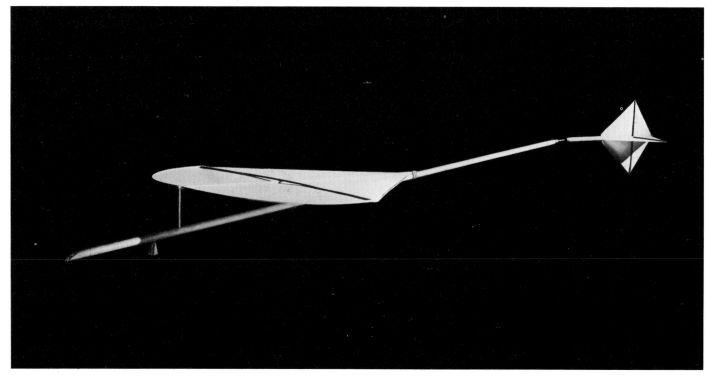

with oscillating wings, stubbornly refusing to accept the impossibility of muscle-powered flight. For example, the French General Resnier de Goué at the age of seventy-two assembled an ornithopter based on an antiquated design and in 1788 dived off the ramparts of Angoulême into the river. Fortunately, he did not injure himself on this occasion, but on making a subsequent attempt broke a leg. A clockmaker from Vienna, Jakob Degen (1756–1846), constructed a machine which was half-ornithopter and half-balloon. The two umbrella-like flappable wings were 130 square feet in area, and were ingeniously designed to allow silk bands to open and close in the same way as the primary feathers of birds. Assisted by the balloon which actually supported over half the total weight and pushing with all his strength, he was able to rise off the ground. In spite of using balloons in a number of public demonstrations, he managed to gain a wide reputation for flying unaided, but in the end he was attacked and injured by an irate and disappointed crowd in Paris. An even more bizarre ornithopter was the steam flapper designed by an Englishman by the name of F. D. Artingstall in about 1830. Encouraged by the progress of railway steam engines, Artingstall built a full-sized steam-powered ornithopter. For its first trial he suspended it from the ceiling, but it flapped so violently that it shook itself to pieces, finally exploding the boiler. Unperturbed by this, he constructed a second machine even more preposterous than the first – this time with four wings flapping alternately like a gigantic dragonfly. An explosion put an abrupt end to this experiment too.

Fortunately for the future of aviation, a small band of optimistic enthusiasts still believed in its ultimate success, and here Cayley's published researches played a major role in guiding them to victory. The most notable of Cayley's admirers was an English engineer, William Samuel Henson (1812–88), who described Cayley as the 'father of aerial navigation'. In 1842 Henson filed a patent for one of the most remarkable and influential aeroplanes in history, the *Aerial Steam Carriage*, the first design for a fixed-wing airscrew-propelled monoplane. Although the machine was never built, it was years ahead of its time, its design owing much to Cayley's influence. It was planned to have double cambered wings made up with spars and ribs covered with fabric (not to become standard practice until 1908) and the wings were to be braced with wires and kingposts. The tailplane had horizontal and vertical movable surfaces to act as elevator and rudder. Henson intended the machine to have a wing-span of 150 feet, huge for those days, and a 30 hp steam engine driving two 6-bladed propellors. This extraordinarily prophetic machine incorporated all the facts of aerodynamics then known, except wing dihedral, and many features which were not to be developed or tested until much later in the century. The aircraft was not expected to take off unassisted, but was to be launched by being sent down a ramp to acquire the necessary initial speed. The chief difficulty which Henson faced was the lack of a suitably light engine of sufficient power – a problem common to all aeronautical pioneers for the next sixty years. If a viable power unit had been available, then the *Aerial* might well have flown.

Undeterred by the lack of power, Henson with the assistance of his friend, John Stringfellow, built a twenty-foot scale model of the *Aerial* to verify the soundness of the

By permission of the Patentees,

THIS ENGRAVING of the FIRST CARRIAGE, the "ARIEL",

is respectfully inscribed to the Directors of

THE AERIAL TRANSIT COMPANY.

basic design. The aircraft was fitted with a small steam engine of Henson's design and the first test was made at Chard, Somerset, in 1845. The machine was launched down a ramp, but due to the excessive weight of the engine, together with the inadequate lift of the wings, it was unable to sustain itself, performing what was sarcastically described at the time as a 'powered glide'. In spite of this disappointment, both men still had confidence in the *Aerial*, but they lacked the necessary money to develop their ideas or build a full-sized machine. Consequently, they employed publicity agents and businessmen to back their enterprise. Unfortunately, these backers were carried away by the project and even tried to promote a bill in Parliament for the formation of an international airline, the Aerial Transit Company, to operate air services to all corners of the globe. One publicity pamphlet was optimistically entitled, 'Full particulars of the Aerial Steam Carriage which is intended to convey passengers, troops, and government despatches to China and India in a few days.' Patent drawings, romantic engravings, prints and souvenirs of the *Aerial Steam Carriage* appeared all over the world, but the

P74 Henson's *Aerial Steam Carriage*, the first propellor-driven aeroplane design, from a contemporary engraving, 1842

enormous publicity had a reverse effect and only succeeded in heaping ridicule on the over-ambitious scheme, and the whole project collapsed. By this time, Henson, discouraged by his failures and unaware of the value of his efforts, had emigrated to Texas.

Stringfellow continued to develop Henson's light steam engine and fitted it into various models based on Cayley's multiplane design, but none of them made a sustained free flight. For several years Stringfellow abandoned aeronautics, but in June 1868 he displayed a model steam-powered aircraft at Crystal Palace, London. It was of triplane construction and clearly of Cayley-Henson ancestry, but when tested proved a failure. This turned out, however, to be Stringfellow's greatest contribution to aviation because it was this machine which persuaded the Wright brothers to adopt

P75 Henson's *Aerial Steam Carriage,* as illustrated on a contemporary tea-towel, 1842

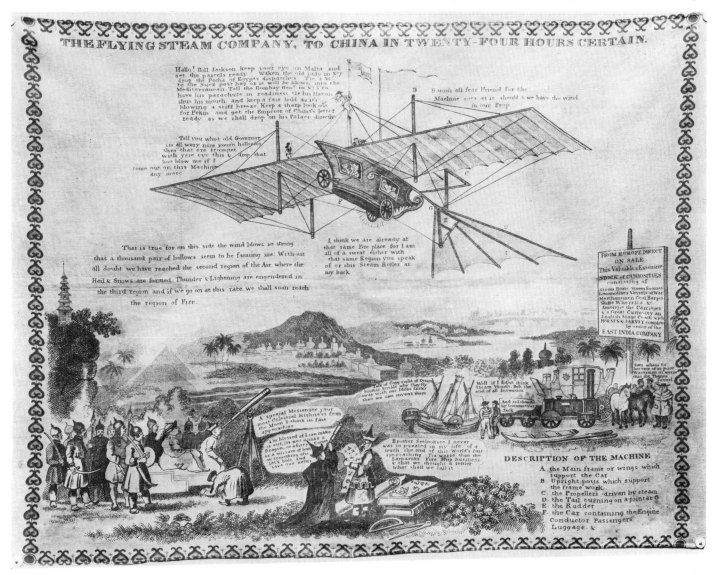

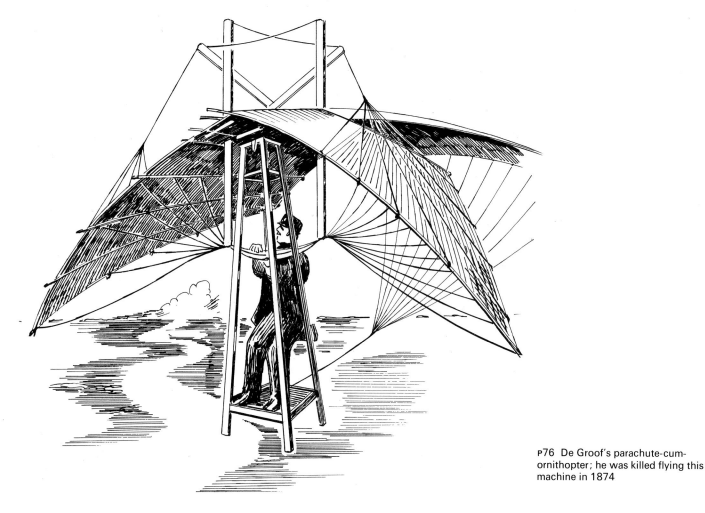

P76 De Groof's parachute-cum-ornithopter; he was killed flying this machine in 1874

superimposed wing configuration in their experiments at the beginning of the twentieth century. Although neither Henson nor Stringfellow completed any successful research or offered anything new to the science of aviation, they showed the world for the first time the shape of the aeroplane to come with its fixed wings and airscrew propulsion. The *Aerial Steam Carriage* fired the imagination of people everywhere and its impact was felt throughout the rest of the century.

After all the excitement that accompanied Henson and Stringfellow there followed a period during which were constructed a variety of flying machines, both model and full-sized, most of which were of indifferent design and had little influence on the future of the aeroplane. The flappers also still lingered on with their misguided and fruitless experiments. For instance, in 1874 a Belgian shoemaker, Vincent de Groof, constructed a hand-and-foot operated parachute-cum-ornithopter, which vaguely resembled one of Leonardo's designs. The operator stood within an upright framework rather like a step-ladder to which the wings and a twenty-foot tail were attached. The machine was expected to fly or glide down to earth after being raised from beneath by a balloon, but when cut loose both wings and tail collapsed due to the air pressure, and de Groof plummeted to his death. A motley assortment of other unorthodox ornithopters appeared during the '60s and '70s, some powered by muscle and others by steam or even elastic. Those ornithopterists, trying to substitute mechanical power for the inadequate muscular energy of the human body, were further frustrated, still believing that the reaction of a bird's downbeat was the total force counteracting its weight. They refused to relinquish this notion in spite of the evidence of the almost motionless wings of gliding birds. Some ornithopters, such as Bourcart's (1863) and Prigent's (1871), were intended to imitate the two pairs of beating wings of a dragonfly. Inevitably they

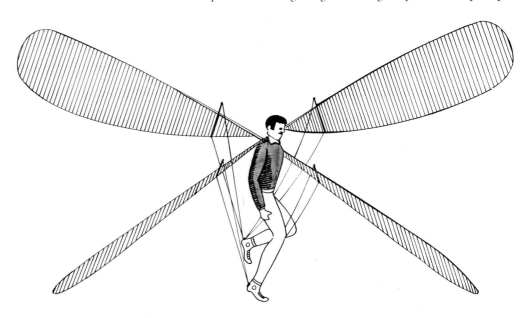

P77 Bourcart's ornithopter, 1863, from a contemporary print

failed, since quite apart from the relationship of power to weight, little enough was then known about the properties of steady airflow, let alone the complex behaviour of unsteady air currents generated by oscillating wings.

European initiatives

By now most of the aeronautical initiative had already shifted to the continent of Europe, and it was here that the first model aeroplane took off and landed under its own power. It was built by a French naval officer, Félix du Temple, in about 1857. A clockwork motor was used in early tests, but this was subsequently replaced by a small steam engine. In 1874 du Temple constructed a full-sized machine of similar design incorporating wings swept forward and set at a dihedral, as well as a rudder, tailplane, retractable undercarriage and a hot-air or steam-driven airscrew. When tested, with a French sailor at the controls, the craft leapt a short distance into the air after a take-off down a ramp. Ten years later another short hop was made in Russia in an aircraft designed by a Russian engineer, Alexander Fedovorich Mozhaiski. It too was made to run down an inclined surface before becoming fleetingly airborne, but in view of the assisted take-offs and the short distances covered, neither of these attempts can be regarded as proper powered flights.

Perhaps the greatest pioneer of French aviation was Alphonse Pénaud, a brilliant figure who tragically committed suicide in 1880 at the age of thirty after his ambitions had been thwarted. He began experiments with model helicopters, but later abandoned them in favour of fixed-wing models which he powered by twisted rubber. One of his machines driven by elastic, the *Planophore*, deserves particular recognition because it was the first to be given inherent stability by wing-tip set at a dihedral and a tailplane set at a negative angle to the wings. The model made a public appearance in the Tuileries Gardens, Paris, and flew up to 131 feet.

During his short like Pénaud pushed the frontiers of knowledge forward by establishing both the theory and practice of stability; but his crowning achievement was an imaginative design for an amphibious monoplane which was patented in 1876, but never built. It was to have cambered elliptical wings set at a dihedral, rear tail unit with rudder and elevators controlled by a single-control column, counter-rotating twin propellors, glass-domed cockpit and a retractable undercarriage with shock absorbers. The aircraft was even to be provided with flight instruments, such as compass and altimeter. Although the design was years ahead of its time, Pénaud was never to put his far-sighted ideas into practice.

For the next quarter century the concept of the inherently stable aircraft became an obsession with European aeronautical thinkers. All round inherent stability is essential for model aeroplanes if they are to fly without a controller, but in full-sized machines which have to be 'flown' by a pilot, too much built-in stability makes them difficult to control and manoeuvre. During the period that followed Pénaud the only aircraft to fly were models which relied on their inherent stability, but when transferred to full-size machines this built-in 'aversion to change' was mistakenly, but understandably,

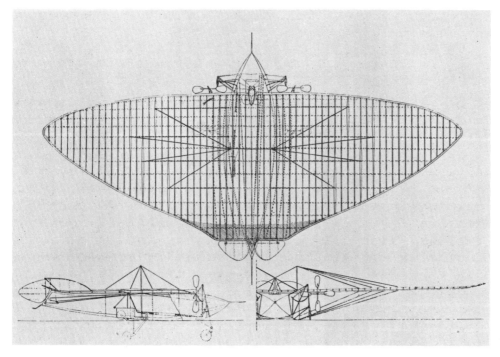

P78 Pénaud's patent drawings for an amphibious monoplane, 1876

overdone, making for insensitive and unresponsive aircraft. This was to remain a stumbling block in Europe until the Wright brothers in America overcame it some twenty years later. The whole problem of stability and control in the air is a complex one, involving not only the basic design of the aeroplane, with its attendant control surfaces and mechanisms, but also the essential human need to master the art of control once airborne.

Birds' wings as models

There were still two other major obstacles to powered winged flight: not only was there the old and frustrating question of power units (lightweight engines of sufficient power were still unattainable), but there was also the problem of the lifting properties of wings, which in those days were inefficient. In 1866 Francis Wenham's (1824–1908) paper, 'Aerial locomotion and the laws by which heavy bodies are impelled through the air and sustained', was read before the Aeronautical Society of Great Britain. It was the first significant milestone in aeronautical theory since Cayley. Wenham's theoretical work was largely based on observations of bird flight which he made on a voyage up the Nile. He not only endorsed many of Cayley's views, but established the fact that the lift derived from a rapidly moving inclined surface is at its greatest near the leading edge, concluding that a long narrow wing (i.e., high aspect ratio) would provide more lifting force than a short stubby one. Furthermore, he pointed out that all birds' wings are cambered and are at their thickest in front where maximum lift is derived. The 'whole secret', he wrote in his paper, 'and success in flight depends upon a proper concave form of the supporting surface.' Before he died at the age of eighty-four, Wenham witnessed man's conquest of the air.

These theories of the greater lifting power of the cambered wing were followed up later in the century by another Englishman, Horatio Phillips (1845–1924), who provided the scientific basis for all modern aerofoil sections. In his laboratory experiments he constructed and used one of the first wind tunnels ever made, and was able to watch the behaviour of a jet of steam as it passed over surfaces of different curvatures. He accurately assessed the ratio of lift to drag produced by a wide range of aerofoils at various angles of attack. He proved that if a wing is curved more on its upper surface than on its lower, most of the lift generated is the result of the reduced pressure above, rather than the positive pressure beneath, as suspected by Cayley. Phillips was, however, unable to explain this behaviour by Bernouilli's principle (see page 17). He imagined that a partial vacuum was created above due to the air bouncing off the leading edge. Nevertheless, his work had a powerful influence on the future of aviation.

If one considers the evolution of birds and insects, it seems both natural and logical that the key to human flight lies in the fixed wing. Perhaps if man had been less preoccupied with the search and application of a prime mover, and been quicker to realize the potential of the motionless wings of gliding and soaring birds, he might have become airborne very much sooner. But it was not until the latter half of the nineteenth century that his efforts were redirected, when once again the bird, that unfailing source of inspiration, encouraged him to take the right course. In a book entitled *Du Vol des Oiseaux*, published in 1864 in France, Ferdinand d'Esterno (1806–83) suggested that man might try harnessing the wind alone for his motive power. The author was one of the most influential of those who advocated man's imitation of the soaring flight of birds, remarking, 'Whoever has seen large birds of prey sailing on the wind, knows that without one flap of their wings they direct themselves as they choose.' But by reasoning that birds in flight shift their centre of gravity in some unaccountable way, d'Esterno seems to have misunderstood how birds effect lateral control. Therefore, when he went on to discuss his design for a glider, he suggested that the pilot control the machine's centre of gravity by adjusting his position in a sliding seat. (A similar technique of flight control whereby the pilot's body was swung about to alter the centre of gravity was adopted by Otto Lilienthal thirty years later, and is still used for controlling modern hang-gliders.) When d'Esterno's proposals were first published, they were generally ridiculed. By the time they had won a certain acceptance and arrangements had been made for the construction of his glider, he was seventy-seven and he died having never seen his ideas put to the test.

The general scepticism with which d'Esterno's theories were greeted was hardly surprising, considering that the finest displays of the soaring and gliding flight of birds were geographically beyond the reach of those most concerned with aeronautics. Gulls and hawks could be seen in Europe, but the true masters of gliding and soaring, such as the albatross, vulture and kite, are mainly to be found in the southern hemisphere. But there were naturally some witnesses to such avian performances. Charles Darwin, for instance, with his perceptive eye was so fascinated by the condors of Peru that in his book, *The Voyage of the Beagle*, he commented, 'Except when rising from the ground, I do not recollect ever having seen one of these birds flap its wings. Near Lima, I watched several for nearly half an hour, without once taking off my eyes: they moved in large curves, sweeping in circles, descending and ascending without giving a single flap . . . and the extended wings seemed to form the fulcrum on which the movements of the neck, body and tail acted. If the bird wished to descend, the wings were for a moment collapsed; and when again expanded with an altered inclination the momentum gained by the rapid descent seemed to urge the bird upwards with the even and steady movement of a paper kite. In the case of any bird soaring, its motion must be sufficiently rapid, so that the action of the inclined surface of its body on the atmosphere may counterbalance its gravity. The force to keep up the momentum of a body moving in a horizontal plane in the air (in which there is so little friction) cannot be great, and this force is all that is wanted.'

The supreme gliding technique of the albatross enchanted the French sea-captain, Jean-Marie Le Bris (1808–72), on his frequent sea voyages to South America. After making careful studies of the bird, he became determined to imitate it and he set about building a full-sized glider based on the albatross. The flexible wings were constructed from wood and flannel and shaped as far as possible like that of the living bird, and the adjustable tail was operated by a foot pedal to steer the machine in the air. For its first test flight in 1857 Le Bris chose to launch his synthetic albatross from a horse-drawn

P79 Le Bris' *Albatross*; a modern reconstruction

cart. After pointing his creation into the wind and signalling to the driver to urge the horse into a gallop, the rope holding the machine down was released and, much to the wonderment of the onlookers, the *Albatross* with its inventor at the controls rose into the air. At a height of 300 feet the triumphant aviator suddenly looked down to see the driver of the cart dangling below him, entangled in the rope. After making a gradual descent, Le Bris managed to make a safe landing for both himself and the terrified man. Although Le Bris made later attempts at becoming airborne, he was never able to repeat his first success; possibly the weight of the dangling man provided exactly the necessary balance to ensure equilibrium. Le Bris appears to have had no scientific training, but he stands along in his early struggles to imitate the majestic flight of these great birds.

Another Frenchman intrigued by the flight of birds was Louis Pierre Mouillard (1834–97), who began his researches in Algeria when he was still a boy. In 1881 he published a book on bird flight, *L'Empire de l'Air*, which became a major source of inspiration to future pioneers of gliding flight. He wrote, 'I hold that in the flight of soaring birds (the vultures, the eagles and other birds which fly without flapping) ascension is produced by the skilful use of the force of the wind, and the steering, in any direction is the result of skilful manoeuvres; so that by a moderate wind a man can, with an aeroplane, unprovided with any motor whatsoever, rise up into the air and direct himself at will, even against the wind itself.' Mouillard believed that wings modelled on the vulture's broad wings could be made to simulate its soaring flight, but although he built a number of machines, none of his experiments were successful. It was the enthusiasm of his writing, rather than any practical experiments, which made the greatest contribution to the art of flying.

Gliding flight

One man particularly fired by the fervour of Mouillard's writing was a German engineer, Otto Lilienthal (1848–96), who was one of the most important men in the history of practical flying. Lilienthal was the first to fly consistently long distances in a heavier-than-air craft. He was unquestionably the most successful pioneer of gliding flight in the world, and his knowledge and experience was to lead to the success of powered flight. After studying the work of his predecessors, Lilienthal made up his mind to study bird flight more intimately than it ever had been done before. He concentrated on the aerodynamics of various types of wings and the relationship between wing area and lift. Although Cayley discovered the basic principles, Lilienthal was the first really to understand how birds propel themselves through the air by means of the airscrew action of their primaries. In 1889 after eighteen years of study and experiment, he published a book, *Der Vogelflug als Grundlage der Fliegekunst*, considering bird flight as the basis of aviation. It remains to this day one of the great classics of aeronautical literature. Within a year of its publication Lilienthal became convinced that no amount of theorizing could ever provide him with all he needed to know about the air, and that the only way to become, in his words, on 'intimate terms with the wind', was to put his ideas to the practical test. Thus began his first tentative efforts at gliding.

During the next seven years up to his death in 1896, Lilienthal experimented with various monoplane and biplane gliders. All were made from tough cotton cloth supported on radiating ribs of peeled willow rods and resembled giant bats. At one stage he made a glider with twisting slots at its wing-tips, which were meant to imitate the propellor action of birds' primaries. The power was to be provided by a small carbonic acid gas engine, but he never did attempt any powered flights. His machines were all hang-gliders, in which the pilot supported himself by his arms leaving his hips and legs free to swing around in any desired direction, thereby enabling control in pitch, roll and yaw by a shift in the position of the centre of gravity. Lilienthal's first gliding experiments were made in the early 1890s from a springboard in his garden, but he soon abandoned this in favour of hill-launching. With experience he learnt to use his body to maintain equilibrium and control in variable wind currents, and this gradually allowed

P80 The model of Lilienthal's glider in London's Science Museum

P81 One of Lilienthal's last flights in his glider

him to increase the height of his launches and the distances flown. In five years of intensive gliding activity (1891–96) he made some 2,500 glides, with controlled flights of up to 1,150 feet. With the steady improvement in the design and technique of his gliders he planned to devote more time to powered flight by means of ornithoptering wing-tips.

About a year before his death Lilienthal began to consider alternative ways of controlling the aircraft in flight. In a letter to a friend he outlined his thoughts about wing warping, steering air-brakes on wing-tips and primitive rudder control. In the last few months of his life he tried to evolve a type of body harness for raising and lowering the tail of the gliders, and it was while experimenting with this that he met with his accident. On a summer's day in 1896 he was gliding in his favourite machine, when a gust of wind suddenly blew him to a standstill in mid-air. He threw his body forward in an effort to get the nose down and pick up speed, but it was too late: the machine stalled and dived to the ground. He died of his injuries the following day in a Berlin clinic.

Lilienthal's influence on the future of aeronautics was profound. At the time of his death he was beginning to arrive at a practical method of flight-control, and had he lived longer, he would surely have accomplished some form of powered flight before the Wright brothers. Moreover, the advances made in photography and printing in the last ten years of Lilienthal's life gave his work an extra dimension. The dry plate negative had only recently been invented by Dr Richard Maddox in 1871, which, apart from being more convenient to use, was fast enough to capture movement. The plate camera had also become a relatively sophisticated instrument and considerable progress had been made in the production of half-tone printing. All this meant that Lilienthal's gliding activities were reproduced in a magnificent series of photographs all over the world; for the first time people could see for themselves that, without an engine and using the natural forces of gravity and wind, man could fly. Perhaps the finest monument to Lilienthal has been the growing band of enthusiasts who, within the last few years, have been drawn to the exciting new sport of hang-gliding.

In the last decade of the nineteenth century power-driven flight was nearly a reality. In England Percy Pilcher (1866–99), as a direct result of Lilienthal's inspiration, was carrying out successful glides in machines of his own design, and had even constructed an engine for powered flight before his untimely death in a gliding accident. In France Clément Ader (1841–1925) constructed a power-driven aircraft looking like a bat, which was reputed to have made a short hop after taking off from level ground. But the immediate future of aviation lay elsewhere: after all the tantalizing slow years of aeronautical progress in Europe, the final stages in the preparation and consummation of man's first sustained powered flight took place on the other side of the Atlantic.

America and final success

Professor Samuel Pierpoint Langley (1834–1906), an American engineer, scientist and astronomer, had already successfully flown models of steam and petrol-driven aircraft for up to three-quarters of a mile before turning his attention to full-sized machines. In 1898 he was asked by the United States Government to develop aircraft for military purposes, but when in due course his man-carrying aeroplane was tested, it proved disastrous: on each of its two trials over the Potomac River, it fouled its catapult-launching mechanism and fell into the water. Even if the launch had been successful, it is highly unlikely that the aircraft would have flown since the volunteer pilot was totally inexperienced; he had never flown in a glider, let alone in a completely untested powered machine. Langley failed to understand that successful flight depends as much on the pilot as on the qualities of the aircraft itself. In view of this failure the war department lost interest in him and withdrew its support, leaving him heartbroken. Because of his misconceptions about flying, Langley made little impression on the science of aviation, but his enthusiasm and confidence in the future of powered aeroplanes played an important part in persuading the Wright brothers to take up flying.

An even more outstanding aeronautical figure in America was the French-born civil engineer, Octave Chanute (1832–1910). His interest in aviation began well before Lilienthal's gliding experiments in Germany, and he carried out in his home town of Chicago an exhaustive investigation into heavier-than-air flight, collecting every available item of information. In 1894 Chanute published a series of articles which were reprinted as a book called *Progress in Flying Machines*, which was to stimulate the Wright brothers to start their experiments with heavier-than-air flight. Chanute thought that the underlying cause of the growing number of failures in the air, culminating in the deaths of both Lilienthal and Pilcher, was the lack of stability. He, therefore, set about designing a glider with a high degree of stability in which the equilibrium was automatically regained after being upset by gusty wind conditions. Although too old to fly himself, he rejected experiments with models, believing them to be inconsistent in open air and 'unable to relate the vicissitudes which they encountered'. For these reasons, he secured the help of a young engineer and conducted tests with full-sized gliders. His most successful machine was a biplane, a direct descendant of Stringfellow's triplane model which was displayed at Crystal Palace in 1868, and in some respects resembled Lilienthal's biplane hang-gliders. It was structurally far superior to anything yet made and certainly helped the Wrights to adopt a biplane construction of similar configuration and rigging. Chanute was a great collector and disseminator of aeronautical information to both sides of the Atlantic; he also took great pains to persuade others to experiment with building and flying aeroplanes. His close friendship with the Wright brothers undoubtedly did much to encourage and support them in their final conquest of the air.

It seems hard to believe that during all this intense activity with heavier-than-air flying machines at a time when man was at the threshold of success, balloons and airships were still the main centre of aeronautical interest. Both the general public and the so-called experts in the field still considered the aeroplane a long way off. Even the distinguished scientist, Lord Kelvin, declared in 1896, 'I have not the smallest molecule of faith in aerial navigation other than ballooning.' He was not to know how rapidly events were to prove him wrong.

Wilbur (1867–1912) and Orville (1871–1948) Wright had been interested in flight since childhood, but their serious study of aeronautics did not begin until after the death of Lilienthal in 1896. Until then the brothers had been engaged in newspaper production and the manufacture of bicycles in Dayton, Ohio, and the money and skills acquired from their business enabled them to start on their new venture. In the early stages the subject of flight was only a hobby to them, but they were soon to become totally absorbed. They read all the available literature, including the works of Mouillard, Lilienthal and Chanute, and before long they arrived at some important new conclusions. They recognized that there were two different prevailing attitudes towards aviation: the first arising from those who made model aeroplanes and the second from those who actually took to the air themselves. In considering their models as functioning independently and separately, the model makers tended to neglect questions of control and manoeuvrability in favour of an over-emphasis on inherent stability. This approach suited model aircraft for which complete stability was a prerequisite of flight, but held back man's own attempts to become airborne. The aviators, on the other hand, designed and flew their aeroplanes with themselves functioning as the central control. They looked on the aeroplane as a means of learning and experiencing flying, but unfortunately put their lives at continual risk: both Lilienthal and Pilcher were such examples. The Wrights, therefore, determined to avoid the mistakes of the one by finding more efficient ways of maintaining control in the air, and the dangers of the other by gaining flying experience without loss of life.

Like many of their predecessors, the Wrights watched birds to see how they managed the fluctuating air currents. Wilbur's observations of the soaring flight of the turkey vulture provided him with a clue – he noted that when it was laterally displaced by a gust of wind, it righted itself by a torsion, or helical twisting, of its wing-tips; if the leading edge of one tip was twisted downward, that of the opposite wing had an upward turn (P51). Perhaps, they wondered, if the wings of an aircraft could be made to twist in a similar manner, the pilot could control and stabilize the aircraft thus, rather than by shifting his body. To achieve this wings would have to be light enough to be 'warped', as they described it, and yet strong enough to lift the aircraft in the normal way (D73).

After trying out wing-warping on a kite, Wilbur and Orville Wright set about in 1900 making their first glider. As well as having a wing-warping mechanism for control in roll, they incorporated another feature which they used in all their early aircraft: they deliberately made the machine unstable. In designing an aircraft which did not have an inherent tendency to right itself when displaced from straight and level flight, they intended to leave the stability to the skill of the pilot. If a gust of wind caught one wing causing the aircraft to bank to one side, the pilot would have to level the wings by applying wing-warp; but they made a mistake in thinking that the machine could be steered by means of wing-warping alone. The glider was completed in autumn of 1900. It was a biplane of seventeen-foot wing-span, the basic biplane concept having originated from Cayley, Stringfellow and Chanute. It had a forward controllable elevator, and the pilot lay prone to reduce air resistance and to minimize the chance of injury to himself while landing. In fact, only a few piloted glides were made in this machine as it was usually flown as a kite with the controls operated from the ground. The Wrights decided to tether the aircraft during its early trials to gain experience from prolonged flying time. The tests were conducted in the lonely sand dunes around Kitty Hawk on the coast of North Carolina, an area of strong and constant winds. The trials were a great success; not only did the wing-warping device and elevator respond well, but the glider confirmed their belief in inherent instability. The Wrights returned home elated, and more determined than ever to make powered flights. 'When once a machine is under proper control under all conditions', wrote Wilbur in his diaries 'the motor problem will be quickly solved.'

In the following year, 1901, they constructed a new glider (no. 2) with a number of modifications including an increased wing-span of twenty-two feet and an anhedral 'droop'. This time it was launched by two men, and operating as a glider, completed glides of up to 389 feet, but unfortunately it had an alarming tendency to slew round

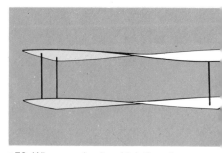

D73 Wing warping in which the wings are twisted to provide lateral control

and sideslip in the direction of positive warp. They also found that the wing camber was too deep, producing an excessive movement in the centre of pressure as the angle of attack was altered. Although some successful glides were made, the brothers were far less pleased than before. They now began to doubt the accuracy of Lilienthal's calculations. 'Having set out with absolute faith in the existing scientific data', declared Wilbur, 'we were driven to doubt one thing after another, till finally, after two years of experiment, we cast it aside, and decided to rely entirely upon our own investigations.'

In 1902 they returned to the drawing board, doing exhaustive research and experiments into all the aerodynamic problems, including the testing of wings in a wind tunnel. The outcome was renewed confidence and their no. 3 glider. Like the no. 2 machine, it was a biplane with the same wing-warping system and anhedral droop, but with a different camber. But the most important difference was the addition of a double fixed fin at the rear, whose purpose was to counteract the previous glider's alarming habit of slipping when warp was applied. In tests the new machine reacted well in straightforward glides, but serious trouble arose while banking. The pilot's efforts to limit the bank by applying positive warp resulted in the wings dropping

P82 The Wright brothers' no. 2 glider being launched into the air at Kitty Hawk with the help of two assistants

further or swinging backwards and causing the aircraft to begin a spin. Before long the Wrights recognized that the central trouble was warp-drag (now equivalent to aileron drag), which produced an increase in the resistance of the wings on one side, and a decrease on the other. In the act of banking the aircraft was slipping sideways through the air in the direction of the bank. The resulting airflow on the side of the fixed vertical fins caused them to act as a lever and rotate the wings about their vertical axis. The problem was finally solved by replacing the vertical fins at the rear by a single movable rudder, with its control cables coupled to the warping mechanism. The warping of the wing-tips was then automatically combined with the rudder so that the rudder always moved in the direction of the bank, thus counteracting warp-drag. In this way, lateral balance could be achieved whenever the aircraft was displaced from the horizontal by the wind, or by the pilot's intentional banking. The new machine made perfect controlled glides in winds up to 35 mph and was able at last to perform smooth banked turns, remaining sensitive and responsive to the lightest touch of the controls. 'The

P83 Orville Wright's historic first flight in the *Flyer I* at Kitty Hawk, North Carolina on 17 December 1903

P84 The model of the Wrights' glider in the Science Museum, London

flights of 1902', wrote Orville, 'demonstrated the efficiency of our system of control for both longitudinal and lateral stability.' The secret of the Wrights' success lay in their concept of stability, and the combination of wing-warping and rudder.

During the autumn of 1902 the two brothers made nearly a thousand glides at Kitty Hawk, becoming experienced and skilled pilots in a comparatively short space of time. At the end of October they returned home once again in high spirits, and immediately started to draw up plans for a larger machine to be propelled this time by mechanical power. The major obstacle was the engine, as suitable ones simply did not exist. These two remarkable men not only designed their own power unit, but partly constructed it. It had four cylinders and was water-cooled, weighing 200 lb. and producing about 12 hp. Since there was no published information available on airscrew design, they were forced to research and develop the design for the propellor.

The Wrights' first powered aeroplane, the *Flyer*, was constructed during the summer of 1903. In appearance and control it was similar to their last glider, except that the rudder and elevators were now double structures. With their new machine the brothers made the now familiar journey to Kitty Hawk. Then came weeks of meticulous preparation and gliding practice with their old no. 3 glider. Finally, on 14 December 1903 after many mechanical breakages and exasperating set-backs, the *Flyer* was ready for its maiden flight. Friends from a nearby coastal life-saving station stood by as witnesses, and the brothers tossed a coin to see who should be the first to fly: Wilbur won. The *Flyer* ran along a take-off rail, climbed steeply, stalled and ploughed into the sand. Wilbur had put on too much elevator, but luckily the damage was only slight and the aircraft was repaired in a couple of days. On Thursday morning of 17

December the weather was perfect with a wind of about 25 mph; it was Orville's turn. The engine was run up and at 10.35 am the restraining rope was released, the *Flyer* gained speed and rose into the air. It flew for 120 feet and landed safely. The flight had only lasted for twelve seconds, but it was, to quote Orville, 'The first in the history of the world in which a machine carrying a man had raised itself by its own power into the air in full flight, had sailed forward without reduction of speed, and had finally landed at a point as high as that from which it started.'

Man had learnt to fly by progressing through stages analogous to those of the insect and the bird millions of years before. He had started with short glides, which gradually led to longer glides and then to flight by powered mechanisms perfected over the years. Throughout the whole process the bird had been man's perfect model and constant inspiration. When Wilbur and Orville Wright were studying the principles of flight, they had frequently returned to the bird to check their theories. 'Learning the secret of flight from a bird', wrote Orville, 'was a good deal like learning the secret of magic from a magician. After you once know the trick and know what to look for, you see things that you did not notice when you did not know exactly what to look for.' By observing the flight of birds man had eventually found the secret which opened up a new era in the history of his world.

5
Aircraft of the twentieth century

THE DAY THAT THE WRIGHT BROTHERS took to the air at Kitty Hawk was the turning point in the history of aviation; but, curiously, it was another six years before anyone on the continent of Europe believed that controlled flight was actually possible. In those intervening years Europeans had no real conception of what manned flight was about; they approached the 'aeroplane' more as a kind of winged automobile to be driven rather than piloted in an open sky. They were still preoccupied with the idea of inherent stability, which kept them from concentrating on such essential matters as flight control.

On 8 August 1908 Wilbur Wright gave a public demonstration of his aircraft near le Mans in France. He was watched by a sceptical crowd of pilots and technicians who stood in stunned amazement as Wilbur put his aircraft through an elegant display of turns, banks and circles with complete mastery and control. The immediate impact on the spectators was profound, but its effect on aeronautics had even more far-reaching consequences, as from that moment European aviation sprung into life. A year later Louis Blériot crossed the channel and the aeroplane became accepted as the world's newest means of locomotion.

Boosted by two world wars and spurred on by commercial and military interests, progress in heavier-than-air flight advanced by leaps and bounds. In little over sixty years after that first demonstration in France man reached the moon. It had taken him the best part of a million years to learn the secret of flight. In his attempts to fly he had struggled with all kinds of inventions from bottles of dew to steam flappers before he was finally rewarded with success. Once a satisfactory formula was found, however, it was used as the basis of the aircraft which we know today.

Structure and function

The fundamental configuration of the conventional aeroplane has in fact changed remarkably little over the years, in spite of the multiplicity of sizes, shapes and designs of modern aircraft. The following features are common to all: one, or sometimes two, pairs of wings supporting a fuselage, tail-unit, undercarriage and engine. Although gliders and sailplanes (the distinction between the two is subtle, the latter being a particularly efficient type of glider capable of thermal soaring) are not power-driven, the great majority of aircraft do have a power unit to provide the driving force. All powered aircraft (excluding those powered by rockets), whether driven by jet or propellor, depend on the backward thrust of air, thereby gaining a reaction or thrust in a forward direction. Although there are various types of jet engine, the principles on which they work are broadly similar. At the front of the engine air is compressed into a combustion chamber by a compressor, where it is mixed with fuel and burnt. With the energy thus acquired, a jet of hot air and burnt fuel is forced out of the rear nozzle at high velocity producing the thrust. Some of the energy released by combustion is absorbed by a turbine wheel which drives the compressor at the front. Most modern sub-sonic jet aircraft, however, now use the more efficient and quieter by-pass or fan-jet. Aircraft powered by propellors, on the other hand, rely on a relatively large diameter jet of unheated air being driven backwards at a relatively low velocity, the blades functioning like rotating wings. The power for the propellor may come from a piston engine or a gas turbine, in which case it is known as a turbo-prop.

All aircraft have to be controlled in the air along three different axes, fixed relative to the aircraft. All three pass through the aircraft's centre of gravity at right angles to one another (D74). The *longitudinal* axis runs from nose to tail, the *lateral* axis runs parallel with a line across the span of the wings, and the *normal*, or *vertical*, axis is at right angles to the other two; thus, the normal axis will be vertical to the ground only when the aircraft is in straight and level flight. Movement about the lateral axis is known as *pitching*, about the longitudinal axis is called *rolling* and about the normal axis is called *yawing*. whereas birds are infinitely flexible in the adjustments they can make to their

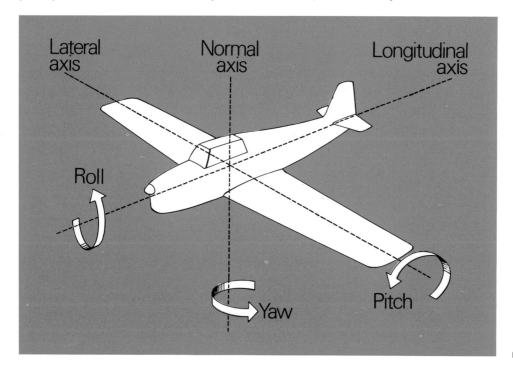

D74 The three axes of movement

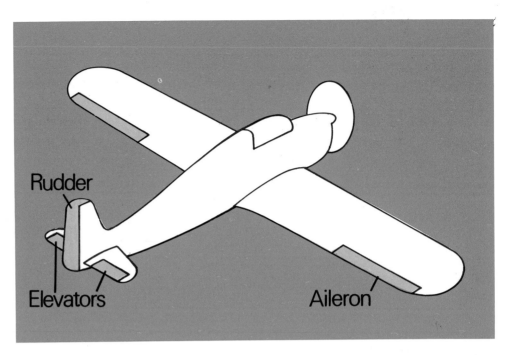

Rudder

Elevators

Aileron

D75 The three principal control surfaces

wings and tail to control their flight, aeroplanes are restricted to the use of three principle control surfaces: *elevators*, *ailerons* and *rudder* (D75). All three are operated by the pilot in the cockpit, the elevators and ailerons by the stick, or control column, and the rudder by the pedals, or rudder bar, on the floor.

Most important are the elevators which are hinged on at the trailing edge of the tailplane and effect longitudinal control, that is along the *pitching* axis. When the stick is moved forwards, they hinge downwards causing the air on their upper surfaces to flow faster and so the pressure above decreases, resulting in the tailplane being 'sucked' upwards. This in turn causes the nose of the aeroplane to pitch downwards (D76).

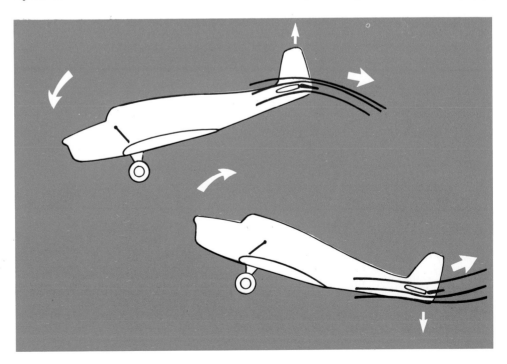

D76 An aircraft's pitch is controlled by the elevators, operated by the stick in the cockpit

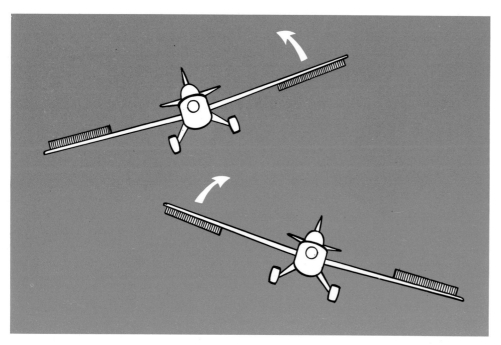

D77 Ailerons, which have replaced the warping system used by the Wrights, control movement along the rolling axis

When the stick is pulled back, the reverse action takes place: the nose pitches up so that the wings' angle of attack becomes greater, increasing lift. Ailerons have now replaced the warping system originally devised by the Wright brothers for controlling movement along the rolling axis. They are hinged on at the trailing edge towards the tip of each wing. When the stick is moved to the left, the aileron on the lefthand wing moves up and the one on the righthand wing moves down (D77). The aeroplane responds by banking to the left owing to the increase in lift on one side and the decrease on the other. The reverse action takes place when the stick is moved to the right. The rudder controls movement along the yawing axis. By depressing either the right or left pedal the rudder

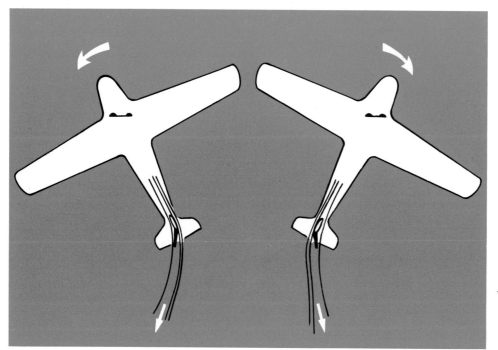

D78 The rudder bar controls the rudder and the aircraft's movement along the yawing axis. To execute a balanced turn the pilot applies aileron and rudder together in the same direction

will hinge accordingly either to the right or left, causing the aircraft's nose to follow in the same direction. The behaviour of the airflow can clearly be seen in diagram 78. As in birds and insects, the movement of these control surfaces at any given angle has far greater effect at high airspeeds than at low. The area of the surfaces and their distance from the axis around which the aeroplane turns also influence their effectiveness.

These three basic movements are known as the *primary* effects of controls, but they also have important and interrelated *secondary* effects. For example, in adjusting the wing's angle of attack, the elevators increase or decrease the drag; because of this, the aircraft's flying speed is also controlled by the elevators. At the same time, by changing the wing's angle of attack the elevators control ascent or descent.

The application of ailerons and rudder also produce secondary effects. Ailerons, which are primarily used to control the bank or roll, induce what is known as *sideslip*. If, for example, the pilot banks his aircraft to the left by moving the stick to the left, weight and lift become out of line and the machine will tend to sideslip towards the direction of the bank (see page 22). The fuselage and fin on the lefthand side then become subjected to a greater airflow, with the result that the aircraft behaves like a weathercock and yaws towards the left, in spite of the rudder being held central. If no action is taken to correct this, an anticlockwise spiral dive will develop. The secondary effect of applying the rudder is to cause the aircraft to roll. When the left rudder is applied, for example, the aircraft will swing towards the left, causing the right wing to move faster than the left (D79). Because the wing with the faster airflow has the greater lift, the machine will bank

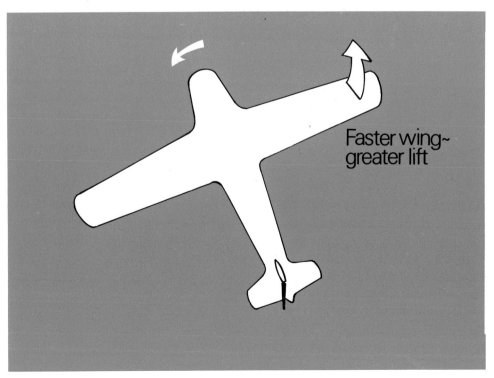

Faster wing~ greater lift

D79 When yawing, the outer wing moves faster than the inner wing, giving greater lift on that side and, a secondary effect, causing the aircraft to roll

or roll to the left, in spite of the stick remaining central. Once again a spiral dive will develop as the nose follows the lower wing downwards. So, although rolling and yawing have their separate controls, the two motions are inseparable. In practice, to execute a normal balanced turn the pilot applies aileron and rudder together in the same direction.

P85 The elevator of a French light aircraft, the Rallye Club, showing the trim tab – the small projecting surface at the rear

Control tabs and trimming

Any change in the weight distribution, whether due to passengers, baggage or fuel, alters the balance of the aircraft. Changes in throttle and airspeed also upset the balance and the unbalanced forces which result have to be corrected by the pilot adjusting his controls. In normal circumstances no great physical exertion is needed to operate the stick or rudder bar, but on a long journey or during an extended upward climb it would be extremely tiring to have to hold the control in the necessary position, especially if conditions were bumpy. The purpose of *trimming* is to reduce the load of the controls, allowing the pilot to fly with 'hands-off' in any attitude. There are a number of methods of achieving this, but the most common is to have a small adjustable surface, known as a *trim-tab*, on the control surface in question (P85, D80). Almost all aircraft are fitted with an elevator trimmer, but generally only the multi-engine types have rudder or aileron trimmers. The elevator trim-tab is mounted on the elevator's trailing edge and can be adjusted by a control wheel in the cockpit in such a way that the airflow over the surfaces of the tab holds the elevator in any given position without the slightest effort on the part of the pilot.

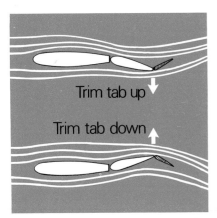

D80 The action of the elevator trim-tab in reducing the load on the stick, allowing the pilot to fly with 'hands off'; the airflow over the surfaces of the tab holds the elevator in any given position

Straight and level flight and speed range

We know that in straight and level flight there is an equilibrium of forces whereby weight is balanced by lift and thrust is balanced by drag. A change in any one of these leads to a change in the others. If, for instance, power is increased by opening up the throttle, the thrust will become greater than the drag, and so the aircraft will accelerate. The resulting increase in speed will generate more lift and the aircraft will climb. The speed of an aeroplane, unlike that of vehicles on the ground, is not primarily controlled by the throttle, but by the elevators. To increase airspeed the angle of attack is reduced by lowering the nose in relation to the horizon. The chief purpose of the throttle is to control the rate of ascent or descent at any particular airspeed, but, of course, throttle and elevators have to be combined in such a way as to achieve the desired effect. For example, if the nose is lowered to increase speed, more throttle will be required to maintain height. If, on the other hand, the speed is reduced by pulling the stick back, the machine will usually climb unless the engine power is decreased.

One of the most important characteristics of an aeroplane is the range of speeds within which it can remain in straight and level flight. Some of the earliest aircraft had hardly any speed range at all owing to their very limited engine power. If they flew at anything less than at full power, they lost altitude. But nowadays there are many aircraft with a maximum speed five times greater than their minimum. To fly at maximum speed and still maintain level flight the pilot has to keep the lift equal to the weight. As the form drag and skin friction resulting from the faster airflow build up, more engine power is required. Eventually a point is reached at full power when maximum speed in level flight has been attained; any further increase in airspeed can only be gained by losing height. To fly as slowly as possible the angle of attack has to be increased to maintain height; the minimum speed for level flight is reached when the induced drag becomes so high that even at maximum throttle the aircraft is unable to maintain altitude. It should be understood that the stalling angle could, of course, be reached before this point. It is interesting to note that aircraft, birds and insects differ from vehicles moving on the ground in that they all need to work at high power in order to move as slowly as possible.

Between the two limits of the speed range there is a speed when an aircraft in level flight flies with least possible thrust and minimum drag; this is at the best lift-to-drag ratio, when the aircraft is flying at the most efficient cruising speed. For jets, however, the most efficient speed can be up to about 30 per cent faster than the speed for minimum drag.

Flaps, slots and stalling

An aircraft's speed range can be considerably extended at its lower end by devices such as *flaps* and *slots*. Flaps were first developed before the First World War, and today come in various forms (D81). They are usually positioned inboard of the ailerons from where they can be extended or withdrawn into the wings' trailing edges (P87). They work by increasing the camber of the wings, and in the case of Fowler flaps there is the benefit of the extra area as well as the slot effect. In some aircraft as, for example, a 747 Jumbo-jet the extent of the flap movement is considerable. When this huge aircraft makes its landing approach, its Fowler flaps can be seen to increase the total area of the wing by about one-quarter.

P86 Fowler flaps fully extended on the wing of a Boeing 707

P87 The rear surface of the wing of a
Boeing 707 with fully-extended Fowler
flaps and, towards the tip of the wing,
ailerons and their trim tabs

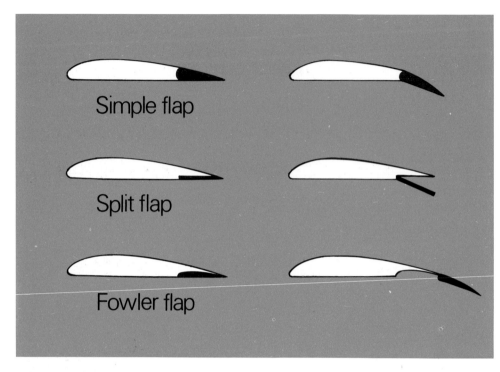

D81 Different types of flaps

Flaps increase both lift and drag and are, therefore, an enormous advantage when landing. By increasing the lift they significantly reduce the stalling speed, thereby permitting lower landing speeds. By increasing the drag they help in two ways: first, they enable the aircraft to descend steeply at the required approach speed allowing for better obstacle clearance (D82). Second, the flap extension makes the aircraft adopt a more nose-down attitude, providing the pilot with better forward visibility. An alternative

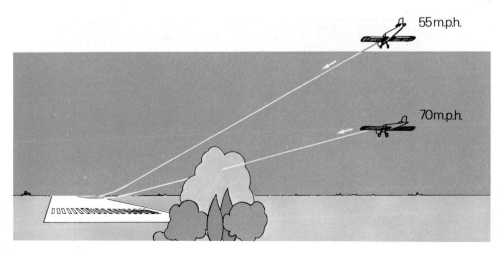

D82 Flaps play a very important part in the landing of an aircraft. They enable the pilot to descend steeply without increasing speed, providing both improved obstacle clearance and better forward visibility

method of steepening the approach path is by increasing artificially the aircraft's drag by the use of *air-brakes* (P88). There are various types and they can appear as flaps, sometimes on the fuselage, but more frequently above or below the wings. Air-brakes differ from slots (described below) in that they neither increase lift nor reduce stalling speed; they have, therefore, no effect on an aircraft's speed range.

P88 The upper surface of a Boeing 707 wing with the air-brakes in the fully raised position

For millions of years birds have been using their slotted wings in slow-speed flying; the British firm of Handley Page Ltd were the first to adopt this principle when they developed slotted wings for their aircraft in 1919. The actual slot is a gradually narrowing gap on the wing's leading edge, and is given its shape by a small auxilliary aerofoil known as a *slat* (D61 page 82 and P91). The slot plays its part by directing a vigorously moving stream of air over the wing's upper surface to reinforce the boundary layer, and so smooths out turbulence and postpones the stall. It can increase the stalling angle to 25° or even 30°, and can sometimes increase the lift by as much as 100 per cent. In light aircraft the slots usually work automatically, lying flush along the wing's leading edge at high speeds and pulled forwards at low speeds as the angle of attack increases. But in larger aircraft the slots are more often linked mechanically with the lowering of the flaps. As well as reducing the stalling speed for a given wing area, the

P89 Leading edge slots closed on a Rallye Club, lying flush along the wing's leading edge

extra lift gained by the use of flaps and slots has also permitted the use of smaller, and so more efficient, wings.

Stalling in one form or another has always been one of the greatest dangers of flying. The danger lies in the loss of height that follows the stall, because it is impossible to recover control until sufficient airspeed has been regained, which may take several hundred feet. The most dangerous and, unfortunately, most frequent stalling occurs shortly after take-off or before landing when there is not enough altitude in which to recover. The development of flaps and slots, however, has led to an enormous improvement in air safety for both large and small aircraft, permitting not only lower landing speeds, but also sometimes eliminating the conventional stall altogether. Stalling, as described in Chapter 1, occurs at high angles of attack, when the airflow breaks away from the wing's upper surface. In the typical stall the centre of pressure

P90 The leading edge slots open. In light aircraft such as the Rallye Club the slots usually work automatically, being pulled forward at low speeds as the angle of attack increases

moves suddenly backwards, upsetting the aircraft's balance (D83). As a result, the nose drops forwards, the tail pitches up and the aircraft rapidly loses height, in spite of the stick being held fully back. In many aircraft the pilot is warned of an approaching stall by *buffeting*, and so he can take corrective action. Control can be effected quite simply by easing the stick forward to regain flying speed, by applying power and then by bringing the nose back again up to the horizon.

At any given angle of attack an aircraft which is heavily loaded has to fly faster than an empty one of the same type because the wing-loading is greater. Consequently, the stalling angle will be reached at a higher speed. When the effects of g (force of gravity) temporarily increase the wing-loading, as in steep turns and aerobatic manoeuvres, the stalling speed also becomes higher, resulting in a sudden, and often alarming, high-speed stall.

P91 This photograph clearly shows the shape of the slat, and the slot it forms when extended

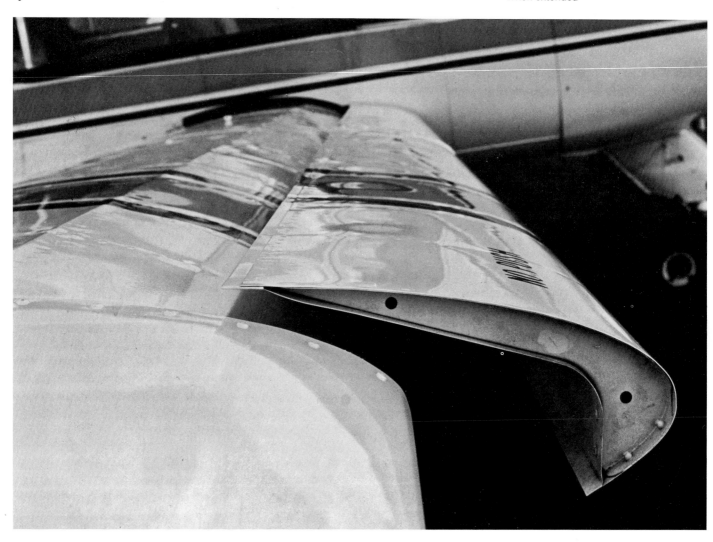

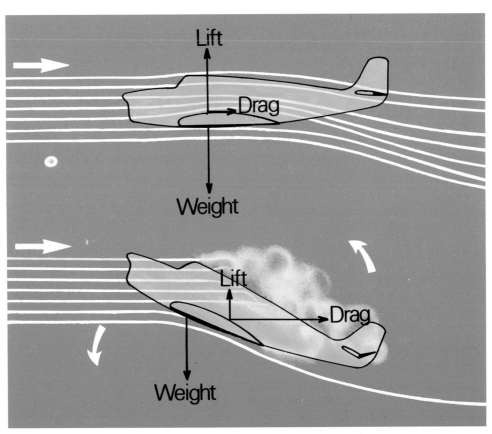

D83 Airflow during normal flight and the stall, showing the change in lift, drag and centre of pressure

For years before it was understood as one of the likely consequences of stalling, the *spin* mystified pilots and caused an enormous number of flying accidents. In the early days aircraft would occasionally enter a spin, apparently without reason, and there seemed little the pilot could do to recover. Fortunately, now that we understand its origin and how to recover from it, the spin is far less dangerous. A spin, or auto-rotation as it is sometimes called, is an unstable stall when the aircraft is simultaneously pitching, yawing and rolling. It is initiated by the onset of a stall, when one wing loses lift before the other. As the wing drops, its angle of attack increases and aggravates its stall. At the same time the ascending wing becomes partly unstalled due to a greater relative airflow from above; the roll and resulting sideslip swing the aircraft into a yaw. If left uncorrected, the nose will drop and follow the fully stalled wing: a spin will develop and continue until the pilot takes suitable action or the machine hits the ground. The pilot's natural reaction is to raise the lower wing by applying opposite aileron; this, however, only makes the condition worse because it tends to increase the angle of attack of the stalled wing even more, adding to drag and so causing more yaw. The only way to recover from a spin is to stop the aircraft yawing by applying full rudder in the opposite direction to the spin, and then to eliminate the stalled condition by easing the stick forward.

The sound barrier

So far we have only considered flight at relatively low speeds, but what happens when flight approaches the speed of sound? Towards the end of the Second World War American pilots flying P-38 Lightnings and English pilots flying Typhoons experienced frightening buffeting when they put their aircraft into steep powered dives at around 600 mph. In some cases the vibrations were so violent that the machines disintegrated: wings and tail-units broke away from the main structure and several pilots lost their lives. The aircraft were in fact moving so fast that the air was unable to move out of the way quickly enough; it became compressed and formed shock waves that hammered on the aircraft's structure.

Up to this time aircraft had only been designed to function at speeds of up to about 500 mph, and little was known of the aerodynamic forces that occurred at higher speeds. As the speed of sound is approached the behaviour of air changes drastically and several of its aerodynamic characteristics work in reverse. A solution to the problem became particularly urgent as it presented a barrier to the higher-speed flight made possible by the new jet engines. There is, and was, no physical 'sound barrier' as such; it was more in the minds of those who thought that the speed of sound was the limiting factor determining man's future progress in the air.

Until comparatively recently it was extremely difficult to study high-speed airflow in the laboratory because wind tunnels became choked at supersonic speeds. The only way to investigate supersonic airflow was to risk actually flying through the barrier. To this end, the United States Air Force ordered a special rocket-powered research aircraft, the Bell X-1, designed to withstand the severe buffeting associated with speeds of around the speed of sound. In October 1947 it was air-launched beneath the fuselage of a B-29 Superfortress at 30,000 feet. After a series of test flights during which speeds were gradually increased, its pilot, Chuck Yeager, finally opened up the throttle to full power. When he approached the critical speed, the machine began to shake alarmingly, nearly going out of control; then, quite unexpectedly, the buffeting stopped – he had passed through the barrier into supersonic flight.

So, what are the characteristics of supersonic airflow which make it so different from ordinary airflow? Although the subject of supersonic aerodynamics is highly technical when considered in detail, a book on flight would be incomplete without at least some explanation, and this simple analogy may help. When a car is driven through a herd of cattle slowly, the cattle saunter to either side in their own time and in an orderly fashion. If the car is driven faster, they will not be persuaded to move out of the way any quicker, as they will bunch up and be unable to adjust their flow pattern to suit the speed of the car. Similarly, when an object is moving through air at around the speed of sound, the air has insufficient time to adjust itself to the new flow. Its molecules are squeezed together and increase in density before they move aside to allow the object to pass.

When an object moves through the air at less than the speed of sound, the air in front of it is warned of the impending disturbance by *pressure waves*. Whether audible or not, these waves are the same as *sound waves*, and are a succession of weak increases and decreases in the density of the air. Travelling ahead of the flying object, the pressure waves warn the air that, for instance, there is lower pressure above the wing than below and that it is, therefore, easier to pass over the top. As a result, the air begins to move aside some distance before the wing hits it, and most of it curves over the wing's top surface. (The warning effect can easily be seen in many of the diagrams of airflow in

P92 Although the Avro Vulcan was not intended to be supersonic, its delta shape influenced the Concorde's design in more ways than one. It also proved an ideal flying test-bed for the Concorde's Olympus engines

Chapter 1.) Pressure waves travel at the same speed as sound, and so, if the aircraft is also flying at the speed of sound, the warning is clearly unable to arrive before the air strikes the machine. This warning is crucial because it influences the behaviour of the air when it meets the aircraft; the difference, therefore, in speed between the aircraft and that of sound is a key factor.

The speed of sound varies according to the temperature of the air: in average temperatures at ground level it is about 760 mph, but in cold temperatures above 36,000 feet it is about 660 mph. The higher the temperature, the faster sound travels. So, from an aerodynamic point of view, the actual speed of the aircraft is not so important as its speed relative to the speed of sound in the particular ambient conditions in which it is travelling. This relationship is expressed in *Mach numbers*. An aircraft flying at a Mach number of 0.5 is moving at half the speed of sound (anywhere from 330 to 380 mph). It is, however, sometimes more convenient to speak in the broader terms of *subsonic*, *sonic* or *supersonic* speeds, rather than specifying the actual Mach number. When an aeroplane moves from subsonic to supersonic speeds, the transition is not instantaneous, but is usually gradual. The range of speeds within which this occurs is known as the *transonic* region. The slow side of the barrier has the more alarming effects on aircraft, and, furthermore, has proved the most difficult obstacle for the designers. Even now less is understood about flight at transonic than supersonic speeds.

As the air has insufficient warning of an aircraft approaching at a transonic speed, it meets the aircraft with a shock, forming a *shock wave*. The shock wave is represented by a line where the relative movement between the aircraft and the air has reached the speed of sound, creating a sudden increase in air pressure, density and temperature (D84). It forms some time before the aircraft reaches Mach 1, and at a point not far from the wing's leading edge where the local airflow is moving at the speed of sound as its velocity rises over the humped surface. The speed at which this first shock wave forms is called the *critical Mach number*; the less slender the aircraft, the lower its critical Mach number. Smaller shock waves may also form on other parts of the machine as, for instance, at the entrance of the jet engine where the local airspeed reaches that of sound. One of the aims of the aircraft designer is to minimize local shock waves and to keep the critical Mach number high so that the transition from subsonic flight is as smooth and rapid as possible.

The aerodynamic forces generated by shock waves are very considerable, and lead to an enormous increase in drag and reduction in lift. Whereas in low-speed flight most of the energy is consumed in overcoming induced drag, at high speeds most of the energy is lost in the form of heat dissipated by shock waves. The airflow behind the shock waves becomes extremely turbulent because the boundary layer separates from the wing's surface, resulting in a loss of lift and a further severe increase in drag. The shock wave and the resulting turbulence produce *shock drag*, increasing the resistance of a conventional aerofoil by many times. Naturally, to offset this a corresponding increase in engine power is needed.

As the speed of the wing increases beyond the critical Mach number, a second shock wave forms on the lower surface and both shock waves move downstream as the

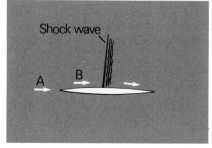

D84 The shock wave: at A the airflow is less than the speed of sound and at B the airflow is increasing to the speed of sound

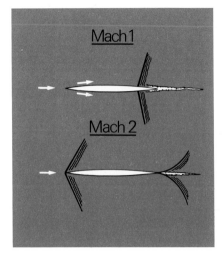

D85 The change in the shock wave pattern with increasing speed

airspeed becomes higher (D85). At Mach 1 the waves attach themselves close to the wing's trailing edge, and when the speed of the aircraft exceeds Mach 1, an additional wave, the *bow wave*, forms in front of the leading edge. Thereafter, apart from the angles becoming more acute with increasing speed, there is little change in the shock wave pattern.

Supersonic flight does inflict certain environmental penalties. Apart from being very expensive in terms of fuel consumption, it brings in its wake the *sonic boom*. We are familiar with sonic bangs in the form of the crack of a whip, the roll of thunder and the report of a high-velocity bullet; they are all caused by shock waves and are the result of a sudden increase in air pressure. When aircraft fly at supersonic speeds, we often hear two shock waves in rapid succession, which correspond to the bow wave and the trailing-edge wave on the wings. It is sometimes possible to hear waves created on the tailplane or smaller ones on other parts of the body, when together they may sound more like a particularly sharp clap of thunder. The intensity of the boom depends on the size, speed and wing-loading of the aircraft; the greater these factors, the more objectionable is the boom. When the aircraft is at close range, the pressure rise in the shock wave can be sufficiently severe to cause damage to windows and buildings.

Having outlined the aerodynamics of supersonic flight, we shall now come back to the effects they have on aircraft design. The early designers were faced with a formidable task. They had learnt that the flying characteristics of aircraft designed for subsonic flight changed dramatically, and even dangerously, as they entered the transonic region. The machines were thrown out of balance by the centre of pressure moving backwards at the onset of the shock wave, which had an alarming tendency to oscillate to and fro on the wing, interfering with the boundary layer and the airflow on the tailplane. The condition is known as a *shock stall*, and it can occur to a lesser or greater degree before the aircraft as a whole has reached the speed of sound. As a result, the machines were subject to violent vibrations and buffeting. Even worse were the effects of shock waves forming on the tailplane, which led to the elevator controls becoming so stiff that they effectively became jammed and, on occasion, reversed the controls of the stick. Unlike the conventional slow-speed stall, however, the shock stall may occur at any angle of attack, although it is most likely at the small angles associated with high speeds. Recovery is made by the use of air-brakes or, if the elevators can be moved, by pulling the nose of the aircraft up to reduce the airspeed.

The designers appreciated the importance of thin streamlined shapes for minimizing drag for speeds of around 400 mph, which were then considered high. As more began to be understood about shock drag and the origin of shock waves, it gradually became clear that aerofoils with low ratios of thickness to chord were essential, and that bulges, bumps and cambered surfaces should be avoided. Complications arose in that the design of aircraft intended for supersonic flight could hardly be less suited to the low speeds essential for take-off and landing. In time the design of supersonic aircraft took on an entirely different form from that of subsonic aircraft. In contrast to airflow at subsonic speeds, which likes smooth rounded contours, airflow at supersonic speeds prefers to meet sharp pointed objects and does not object to having to change direction suddenly.

P93 RAF Phantoms in flight formation. This aircraft has considerable sweepback and is capable of speeds approaching Mach 2 in level flight

P94 The Rallye 180 GT is a popular and easy to fly French light aircraft. To improve the streamlining of the fixed undercarriage it is provided with spats

P95 The VC 10 is perhaps the most beautiful of subsonic passenger aircraft. With the weight of its four engines at the rear, the wings are placed much further back than those aircraft with wing-mounted engines. The leading edge slats can be seen extended

P96 The beautiful gothic shape of the Concorde was the result of 4,000 hours of testing in a wind-tunnel. The aircraft is so aerodynamically efficient that it does not require high-lift devices such as flaps or slots. There can be little doubt that the SST aircraft will lead the world's air routes in the future

P97 The McDonnell Douglas DC-10, shortly after take-off, with flaps and slats still extended

One of the most significant contributions to high-speed flight was the application of *sweepback* (D86). Slightly swept back wings had already been used in aircraft for adding stability and improving the pilot's view from his cockpit, but by substantially increasing the angle of sweepback to 30° or 40° or even more, the critical Mach number of the wing can be raised, delaying the onset of shock stall and sometimes avoiding it altogether. The wing's geometry is so arranged that a lower air velocity is obtained at right angles to the leading edge (across the chord of the wing), for this is the velocity which determines the wing's critical Mach number. Aircraft with swept back wings can be made to fly at much higher speeds than those with relatively straight wings without producing shock stall, allowing for instance, today's jet airliners to fly at speeds approaching Mach 1. Performance at speeds around Mach 1 can also be improved by nipping in the waistline to avoid the bulges which cause shock wave and drag according to the so-called *area rule* (D87). This principle means that the cross-section of the aircraft is kept more or less constant from nose to tail, offsetting the additional area presented by the wings.

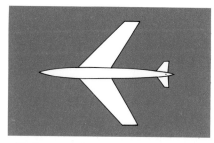

D86 Swept back wings delay the onset of shock stall allowing jet airliners to fly at speeds approaching the speed of sound

The most popular plan form for supersonic flight is the *delta* shape (P92). It has similar high-speed qualities as swept back wings, but it is stronger and provides longer trailing edges on which control surfaces can be mounted. Delta wings also require a higher angle of attack to produce the necessary lift at low speeds than is required by conventional aircraft, as can be seen in the Concorde when it lands and takes off. Thin delta wings, however, present certain aerodynamic complications, because they create patterns of airflow which do not take place with conventional wings: some of the air at the sharp leading edge separates to form a concentrated vortex which flows backwards over the wing's surface, and at the same time the boundary layer at the rear tends to flow outwards. By modifying the plan form of the delta the leading-edge vortex can be persuaded to adhere to each wing without breaking up, even at slow speeds up to and beyond the stall. This is done by making the wing's leading edges curved with rounded tips, and by increasing the sweepback from extended roots (P97). Such a form gives the Concorde its striking Gothic appearance and can improve an aircraft's lift by over one-third – in fact, Concorde is so designed that it does not need flaps or slots at all.

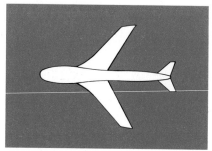

D87 Area rule: nipping in the waistline of an aircraft (exaggerated in the diagram) can improve performance at high subsonic speeds

In common with many other delta wing aircraft, the Concorde does not have a tailplane. Although the tail plays an essential role in the maintenance of stability in conventional aircraft, there are several good reasons for doing away with it in high-speed machines. By eliminating as much as possible that does not contribute to lift, parasitic drag can be kept to the minimum. Without a tail or elevators the aircraft becomes a 'flying wing', which means that some alternative way has to be found for maintaining longitudinal stability and controlling pitch. As any restoring control surface needs to be well behind the centre of gravity, the obvious place for it is on the tips of the swept back delta wings; thus, the wing-tips now act in the same way as a tailplane. To take the place of the elevators the trailing edges of the wings are also provided with *elevons*; these combine the functions of the elevators and ailerons of conventional aircraft.

Concorde's control and stability in the transonic region, which, as already mentioned, is one of the most difficult areas of supersonic flight, is improved by having an adjustable centre of gravity – perhaps one of the few characteristics it shares with the hang-glider. As the centre of pressure moves back at transonic speeds, fuel is pumped from trimming tanks at the front of the wings to others at the back of the machine, where it remains during supersonic cruise. When Concorde slows down to subsonic speeds, the centre of gravity is brought forward again by fuel being pumped back to the front.

Many high-speed aircraft are equipped with small projections of various shapes towards the leading edge of the wing's top surface, which are known as *vortex generators* (P98). They liven up the airflow within the sluggish boundary layer, enabling the air to move further over the wing's surface before separating and becoming turbulent. In this respect, they have a similar effect to leading-edge slots. One method of reducing drag and

P98 Vortex generators on the wing of a Boeing 707. These projections act in a similar way to leading edge slots by livening up the airflow in the boundary layer and keeping turbulence on the wing's upper surface to a minimum

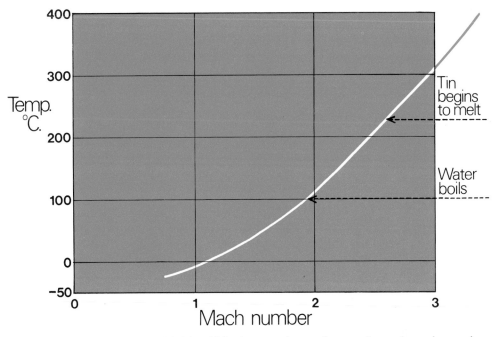

D89 The rise in the skin temperature as the Mach number increases

P99 (*left, above*) Hang-glider on the wing: to initiate a turn to starboard, the pilot has shifted his weight by pushing the control frame to his left

P100 (*left, below*) A hang-glider preparing for take-off. The bar behind the pilot's ankles supports his feet once he becomes airborne, allowing him to fly in the prone position

D88 The principle of the variable-geometry wing was proposed in the 1940s by Dr Barnes Wallis. At low speeds the wings pivot outwards to increase lift while at high speeds the wings assume a delta shape

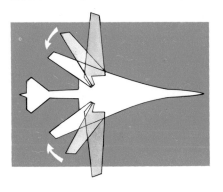

maintaining laminar flow, which is still in the experimental stage, depends on the suction of turbulent air into hundreds of holes on the wing's upper surface. Reducing drag by this means could result in a fuel saving of some 30 per cent.

The versatility and speed range of aircraft intended for flight at high Mach numbers can be improved by imitating the high-speed flight of diving birds, such as falcons and gannets, which fold their wings towards their bodies with the tips swept back during a dive and extend them outwards during take-off and landing. The idea of variable-geometry wings in aircraft was originally proposed in the 1940s by an Englishman, Dr Barnes Wallis. His basic notion was simple: at high speeds the wings assume a delta shape, while at low speeds to provide the extra lift and a sufficiently low induced drag, the wings pivot outwards to increase both the area and aspect ratio (D88). Needless to say, such a system entails a host of practical and mechanical difficulties, such as the strength required by the pivots which have to bear enormous aerodynamic loads and the problem of redistributing the weight to maintain trim as the position of the wings is adjusted.

When aeroplanes fly at Mach numbers of 2 or more, the very high skin friction presents another obstacle in the form of overheating (D89). The problems this brings include finding suitable materials which retain adequate strength at high temperatures, preventing the fuel boiling in the tank and, of course, insulating the passengers and crew from the searing heat. Even if we learn to live with sonic booms, this so-called heat barrier will certainly prove the limiting factor for very high-speed supersonic flight in the atmosphere for several years.

The faster we fly and the more we learn about aeronautical science, the less 'real' to us our aeroplanes seem to become. We have only to look at the atrophied wings of some supersonic aircraft to realize that, with increasing speed, they begin to function more like rockets, relying on engine thrust rather than the lifting properties of wings. Although fling is now a safer activity than ever before, there is no doubt that, like so many other aspects of our life, it is also becoming more impersonal and remote. As passengers are encouraged to listen to piped music, watch films or eat and drink in mid-air and as aircraft become larger and fly faster and higher, the sensation of flight is fading. With such sophisticated electronic aids as auto-pilots, inertial navigation and fully automatic instrument-landing systems, piloting a modern jet is more like driving a computerized aerial bus than flying an aircraft; pilots have to concentrate on instrument panels, check-lists and columns of figures, rather than the air, clouds and sky around them.

Hang-gliders

It is hardly surprising that, in these circumstances, hang-gliding has now become such a popular sport. It is almost as if we have turned full circle, reverting to leaping off hillsides and cliffs on fabric wings like the medieval tower-jumpers. Hang-gliding is the simplest and cheapest form of flying, and is available to anyone who is reasonably fit and bold enough to try. Hang-gliders can be flown from any suitably sloped hill facing the prevailing wind. The sport is quiet, uncomplicated and free: it has few rules or regulations, does not need runways, radios, control towers or winches and leaves behind no traces. Above all, it is more exhilarating and gives a greater feeling of being in harmony with nature than any other form of flying. One reason for this is that the pilot is an essential part of the structure, weighing about four times as much as his 'wings', which become almost an extension of himself. Unlike other flying machines the pilot has a bird's-eye view of the ground beneath as he wheels and soars almost with the grace and freedom of a gull. Perhaps too, like mountaineering, the element of risk adds to its appeal. Although in many respects hang-gliders are easier to fly than light aircraft, they do carry more risk, and in the early stages of learning the sport requires considerable courage and determination.

Experienced pilots with advanced hang-gliders can today achieve spectacular performances which were once thought only possible with powered aircraft and sailplanes. Given the right conditions, pilots can remain airborne for several hours, ascend in thermals to thousands of feet, make long cross-country journeys and even perform certain aerobatic manoeuvres. The remarkable feature is that all this can be accomplished with an 'aircraft' which can be folded up, carried on the shoulder and stored in the garden shed.

Hang-gliding, of course, started with Otto Lilienthal and is the point at which successful manned aviation began. The difference between his attempts and those of his present-day followers is that Lilienthal was not only testing and developing his glider, but was also learning to fly without the benefit of the previous experience of others. Most modern hang-gliders date back to 1951 when an American, Dr Francis Rogallo, patented a delta-wing kite with flexible surfaces, whose shape was maintained by tension wires. As part of a multimillion-dollar research project for the recovery of space capsules, the United States space agency, NASA, developed and thoroughly tested the kite in wind tunnels; it was shown to possess, among other characteristics, a very high level of inherent stability. When the project was subsequently abandoned, several people were quick to realize its possibilities. In the late 1960s two Americans, Michael Markowski, an aeronautical engineer who worked on the design of the Douglas Aircraft Company's DC-10 and Richard Millar, editor of the American magazine, *Soaring*, saw a potential use for the Rogallo sailwing, as it was called, for hang-gliding. They adapted and developed it for foot-launched flight from the tops of sand dunes, and in due course it became refined and improved by the experience of others.

The modern hang-glider has a flexible aerofoil surface, known as the sail, made up from a fabric, such as dacron. This is supported by a tubular aluminium frame, which is strengthened by a number of tension lines. Because the Rogallo hang-glider is so light, simple and stable, it requires no elevators, ailerons or rudder and is controlled by the pilot shifting his weight in relation to the kite. Hang-gliding is subject to exactly the same laws of aerodynamics as any other flying machine. When the angle of attack is zero, the sail lies parallel to the airflow and, obviously, derives no lift. As soon as the angle of attack is increased, the sail fills with air and lift is generated. Directional stability is provided by the keel formed between the two cones of the sail, and the exceptionally high degree of lateral stability is provided by the weight of the pilot hanging underneath giving pendulous stability (P99).

The pilot either sits in a seat or lies in a prone position hanging in a harness attached to the top of a triangular shaped control frame at the centre of gravity. He controls the glider by holding the bottom of the frame and moving his body about. By pulling the frame towards him, the pilot shifts his weight forwards in relation to the glider so that the nose drops and the airspeed is increased. If the pilot wishes to ascend or reduce his speed, he simply pushes on the frame. As these controls are opposite to those of an aeroplane, when pushing forward on the stick causes the machine to pitch down, it can be confusing for aircraft pilots taking up hang-gliding. Turns are also executed by shifting weight: if the pilot moves his body to the left, the shift in weight causes the glider to bank so that the lift acts sideways and the glider starts a left turn.

Although cross-country flights are possible in places where strong thermals are in plentiful supply, in temperate climates most flights are made by riding the up-currents at the tops of hills and cliffs. In the absence of a suitable rising current the pilot has to content himself with a gradual descent to the bottom of the hill, from where he has to make a strenuous climb back up before attempting another flight. The performance of the hang-glider can be much improved by increasing its aspect ratio. The typical Rogallo has a glide ratio of 1:5, which means that for every one foot of descent it will move forward five feet in still air. (Sailplanes, in comparison, can sometimes have a glide ratio of 1:50.) If taken too far, however, higher aspect ratios not only make the hang-gliders more difficult to handle, particularly in high winds, but also require control surfaces, auxilliary aerofoils and slots which detract from the Rogallo's principal attraction of its basic simplicity.

In contrast to conventional aircraft, most hang-gliders have mild stalling characteristics; if they are flown too slowly, they pitch down gently rather than suddenly entering a nose-dive. Perhaps the greatest danger is stalling at intermediate heights of about 25 feet, when the novice may have insufficient altitude in which to recover. Most stalls, unfortunately, occur also at high ground speeds when flying downwind; in these circumstances the inexperienced pilot, seeing the ground moving so rapidly beneath him, has a tendency to reduce his airspeed with catastrophic results – the glider stalls and the unprotected pilot can crash to the ground at over 30 mph. There is little doubt that in the hands of the careless or irresponsible hang-gliding is potentially more dangerous than most other sports such as skiing and mountaineering. Nonetheless, in terms of rewards it ranks exceptionally high, allowing one to experience perhaps the most exhilarating and enjoyable form of flying yet devised by man.

Although man is by his very nature essentially earthbound, there is no longer any need for us to envy insects and birds in quite the way that our forefathers did. With the artificial wings of aircraft we can already outdistance and outpace all other living things. Our first wings, if they may be called such, were as unspecialized in their way as those of the first flying insects or *Archaeopteryx*, but unlike them they evolved in an infinitely shorter period of time. The fragile linen-covered surfaces, on which the first forays into the air were made, have developed within a few years into a bewildering number of forms, from the sleek metal aerofoils which are the products of computers and advanced technology to the simple sails of the Rogallo hang-gliders.

This book has tried to explain the basic principles that underlie all forms of flight, whether by insect, bird or man. The aerodynamics involved in lifting an object off the ground and keeping it in the air are universal, but one of the fascinations of the subject is that there are still aspects of it that remain to be discovered and explored. In man's exploration of flight, both high-speed cine and still photography have played a vital part, not only revealing the movements of wings not observable by the naked eye but also contributing to our final conquest of the air. Now that we have acquired experience of wings we are better able to assess the aerial performance of other flyers and to understand more fully the aerodynamic properties of their bodies. We can look at the wings of insects and birds with greater insight and appreciate the processes which have made each creature perfect for its aerial rôle and at the same time a supreme work of art.

A note on the photography

MY BASIC PHOTOGRAPHIC EQUIPMENT consists of a 35 mm Leicaflex SL with the Leitz 100 Macro Elmar and 135 mm Hector lenses; while the colour film used for almost all the photographs included in this book is Kodachrome 2, an emulsion capable of unequalled quality of image, but which has now unfortunately been discontinued. The equipment employed for taking the high-speed photographs of insects and birds was specially developed for the purpose, for there was no commercially available photographic apparatus capable of satisfying my highly specialized requirements. The insects posed the most serious problems since, apart from being considerably smaller than birds and having much higher wing-beat frequencies, their flight paths are erratic and unpredictable. I had to devise a system for detecting them accurately in a precise plane of space, which then, after the shortest possible delay fired a shutter synchronized with a short duration and powerful electronic flash unit. The principle behind the system finally adopted is quite simple: the insect, or bird, is persuaded to fly through a narrow beam of light focused onto a photo-electric cell which then triggers in sequence an amplifier, a shutter power unit and shutter and flash unit (D90).

The chief difficulty lay in the design of the shutter and flash apparatus. My existing camera shutter was useless, as it, in common with all conventional shutters, had a time delay of about 1/20th second before it opened fully, long enough for an insect to move ten inches away from the plane of focus and out of the picture entirely! Thus, a special shutter was designed which reduced the shutter delay to about 1/400th second. This was clamped in front of the lens and used instead of the Leica's focal plane shutter which was locked fully open.

Another problem was the flash; it had to be of sufficiently short duration to 'freeze'

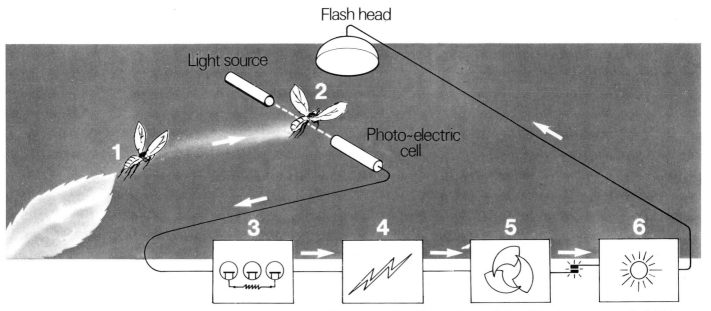

D90 Schematic diagram illustrating the technique used by the author for photographing birds and insects in free flight: 1. Insect takes off; 2. Light beam broken; 3. Amplifier triggered; 4. Shutter power unit triggered; 5. Shutter opens, closing flash contacts; 6. Flash fires

the movement of insects' wings, and yet be powerful enough to expose Kodachrome film at a small aperture for maximum depth of field – two conflicting requirements. As there were no suitable flash units on the market, the apparatus had to be custom-built. After much experimental work over a couple of years, with an electronics engineer colleague, a sufficiently powerful unit was developed which was capable of providing a flash duration of a 1/25,000th second. Furthermore, the equipment was portable, and so could be used outside on location for bird photography. Later, a refinement was added which permitted a sequence of flashes to be fired with an adjustable time delay between each. This enabled me to expose a number of images on the same piece of film, allowing me among other things to study the take-off and landing behaviour of insects and birds. More details about these photographic problems and how they were solved can be found in my book *Borne on the Wind* published in 1975.

Bibliography

BROWN, R. H. J. 'The Flight of Birds'. *Journal of Experimental Biology.* 25, 322–33 and 30, 90–103. 1948.

CHANUTE, O. 'Aerial Navigation'. *Cassiers' Magazine.* Vol. XX, 1901

CLANCY, L. J. *Aerodynamics.* Pitman, London 1975

DALTON, S. N. *Borne on the Wind: the Extraordinary World of Insects in Flight.* Chatto and Windus, London, and Reader's Digest Press, New York 1975

DESMOND, A. J. *The Hot-blooded Dinosaurs.* Blond and Briggs, London 1975

ESTERNO, M. D'. *Du Vol des Oiseaux.* Paris 1864

GIBBS-SMITH, C. H. *Aviation.* HMSO, London 1970

GRAY, J. *Animal Locomotion.* Weidenfeld and Nicolson, London 1968

LIGHTHILL, M. J. 'On the Weis-Fogh Mechanism of Lift Generation'. *Journal of Fluid Mechanics.* 1973

LILIENTHAL, O. *Der Vogelflug als Grundlage der Fliegekunst.* (Bird flight as the basis of aviation) R. Oldenbourg, Berlin 1889

MAREY, E. J. *Le Vol des Oiseaux.* G. Mason, Paris 1890

NACHTIGALL, W. *Insects in Flight.* George Allen and Unwin, London 1974

PENNYCUICK, C. J. *Animal Flight.* Studies in Biology no. 33. E. Arnold, London 1972

PRANDT, L. *Essentials of Fluid Dynamics.* Blackie, London and Glasgow 1952

PRINGLE, J. W. S. *Insect Flight.* Cambridge University Press 1957

WEIS-FOGH, T. 'Quick estimates of flight fitness in hovering animals, including novel mechanisms for lift production'. *Journal of Experimental Biology.* 59, 169–230. 1973

WRIGHT, W. & O. *Miracle of Kitty Hawk.* (The letters of Wilbur and Orville Wright). New York 1951

Acknowledgements

STEPHEN DALTON would like to thank the following for their help with writing this book: Charles Ellington of the Department of Zoology, Cambridge University, for his advice about the aerodynamics of flapping flight, and for allowing him access to his latest researches; Graham Slater for his advice about hang-gliding; Professor C. J. Pennycuick for kindly reading the third chapter; Chris Le Cluse and Grant Bradford for their skilful diagrams; his editor Sarah Riddell for her patient and enthusiastic help; and finally his wife Liz for slaving away at the typewriter in the preparation of the manuscript.

Blacker Calmann Cooper Ltd. would like to thank the following for providing photographs which are reproduced in this book: British Aircraft Corporation (P96), British Airways (P96), the Central Office of Information (P93, P95), the Mansell Collection (P64), and the Science Museum, London (P62, P65, P67, P70, P73–75, P80–83). They would also like to thank J. M. Dent & Sons Ltd and E. P. Dutton & Co Inc for granting permission for including excerpts from the Everyman Library edition of *The Voyage of the Beagle* by Charles Darwin. All the other photographs were taken by the author.

Index

Reference numbers to plates and diagrams are in brackets.